P9-DHD-565

PRAISE FOR

MOST WANTED PARTICLE

An *Observer* Top Ten Science and Technology book

"Most of the existing popular accounts of the events leading up to the July 2012 discovery claim at CERN are written from a theoretical perspective by outsiders. Jon Butterworth is an experimentalist and is **the first to give a vivid account of what the process of discovery was really like for an insider**."

—**Peter Higgs**, Winner of the Nobel Prize in Physics

"The story of the search for the Higgs boson is so **edge-of-your-seat exciting** that it practically tells itself—but still, why not get **the story from someone who was there for every step along the way**? Jon Butterworth is a talented writer and a world expert in the physics, and his book is hard to put down."

—**Sean Carroll**, physicist at Caltech and author of *The Particle at the End of the Universe*

"This is **a unique book**, which captures the highs and lows of the last 20 years of particle physics, culminating with the discovery of the Higgs Boson. I've known Jon for most of my career—he's an insightful, creative, diplomatic and occasionally outspoken physicist, and every facet of his character is on display in this beautifully written book. **If you want to know what being a professional scientist is really like, read it!**"

—**Brian Cox**, author of *Why Does* $E=mc^2$? and *The Quantum Universe*

"If you met Jon Butterworth in a pub—which, judging from the many anecdotes in Most Wanted Particle, is a non-trivial probability—his is the voice you'd like to hear, this is the tale you'd want him to tell: a breezy recounting of the discovery of the Higgs boson that turns out to be both **an accessible primer on particle physics and a lively look at behind-the-scenes Big Science**."

—**Richard Panek**, author of *The 4% Universe: Dark Matter, Dark Energy, and the Race to Discover the Rest of Reality*

"[A] **charming, enlightening** bulletin from one of the most exciting fields of human endeavor."

—*Guardian*

BECAUSE EVERY BOOK IS A TEST OF NEW IDEAS

"The book contains a **fascinating inside perspective of the discovery of the Higgs boson**. It offers an insight into the intense, bewildering and intimidating media scrutiny that physicists aren't used to, combined with intimate details about the life of a high-powered physicist and some lovely explanations of the physics behind the discovery."
—*New Scientist*

"This is more than just another telling of the story of the hunt for the Higgs at the LHC—**the reader here is utterly immersed in the politics, excitement and sheer intellectual adventure of discovery . . . from someone who was actually there!** The process of scientific research is laid bare in all its glory, warts and all, and emerges as a delightful example of what is best about human intellectual endeavor."
—**Jim Al-Khalili**, author of *Quantum: A Guide for the Perplexed*

"Like *The Lord of the Rings, Most Wanted Particle* takes readers on a long path with many moments of peril and uncertainty to reach the triumphant discovery of the Higgs Boson. It is **a great chronicle** of a part of the endless chain of progress in science at the LHC."
—**Jim Gates**, University System of Maryland
Regents Professor of Physics

"A smashing journey."
—*Physics World*

"**An excellent, accessible guide to one of science's greatest discoveries** . . . vivid insights into the doing of science, including the customs of various scientific tribes at CERN."
—*Sunday Times*

"The mix of technical description, anecdote and humour works brilliantly and feels completely fresh in my experience of science writing—it really **unlocks the holy grail of combining entertainment and understanding**."
—*PFILM*

"**Riveting!** Gonzo journalism but in the entrails of experimental particle physics."
—**Pedro G. Ferreira**, author of *The Perfect Theory*

"A great read if you're curious about the Higgs boson, the work done at the LHC, what it's like to be a physicist or how life as a research scientist has to dovetail with the 'real' world in terms of politics, economics and justifying to the public why science is important and should be funded. **If you're remotely curious about the universe, read this**."
—**Steven Thompson** of Physics Steve,
a theoretical physics blog

MOST WANTED PARTICLE

JON BUTTERWORTH

FOREWORD BY LISA RANDALL

NEW YORK

MOST WANTED PARTICLE: *The Inside Story of the Hunt for the Higgs, the Heart of the Future of Physics*

First published in the UK as Smashing Physics by Headline Publishing Group, 2014

The Experiment, LLC
220 East 23rd Street, Suite 301
New York, NY 10010-4674
www.theexperimentpublishing.com

The Experiment's books are available at special discounts when purchased in bulk for premiums and sales promotions as well as for fund-raising or educational use. For details, contact us at info@theexperimentpublishing.com.

Library of Congress Cataloging-in-Publication Data

Butterworth, Jon.
[Smashing physics]
Most wanted particle : the inside story of the hunt for the Higgs, the heart of the future of physics / Jon Butterworth.
pages cm
First published in Great Britain in 2014 as Smashing physics, by Headline Publishing Group.
Includes index.
ISBN 978-1-61519-245-8 (hardcover) -- ISBN 978-1-61519-246-5 (ebook)
1. Butterworth, Jon--Career in physics. 2. European Organization for Nuclear Research. 3. Large Hadron Collider (France and Switzerland) 4. Higgs bosons. I. Title.
QC16.B88A3 2015
539.7092--dc23
[B]

2014036045

ISBN 978-1-61519-245-8
Ebook ISBN 978-1-61519-246-5

Jacket design by Catherine Casalino
Cover graphic depicting ATLAS experiment particle collisions courtesy of CERN
Author photograph © Claudia Marcelloni 2014
Typeset in Berkeley by Avon DataSet Ltd., Bidford-on-Avon, Warwickshire

Manufactured in the United States of America
Distributed by Workman Publishing Company, Inc.
Distributed simultaneously in Canada by Thomas Allen & Son Ltd.

First printing January 2015

10 9 8 7 6 5 4 3 2 1

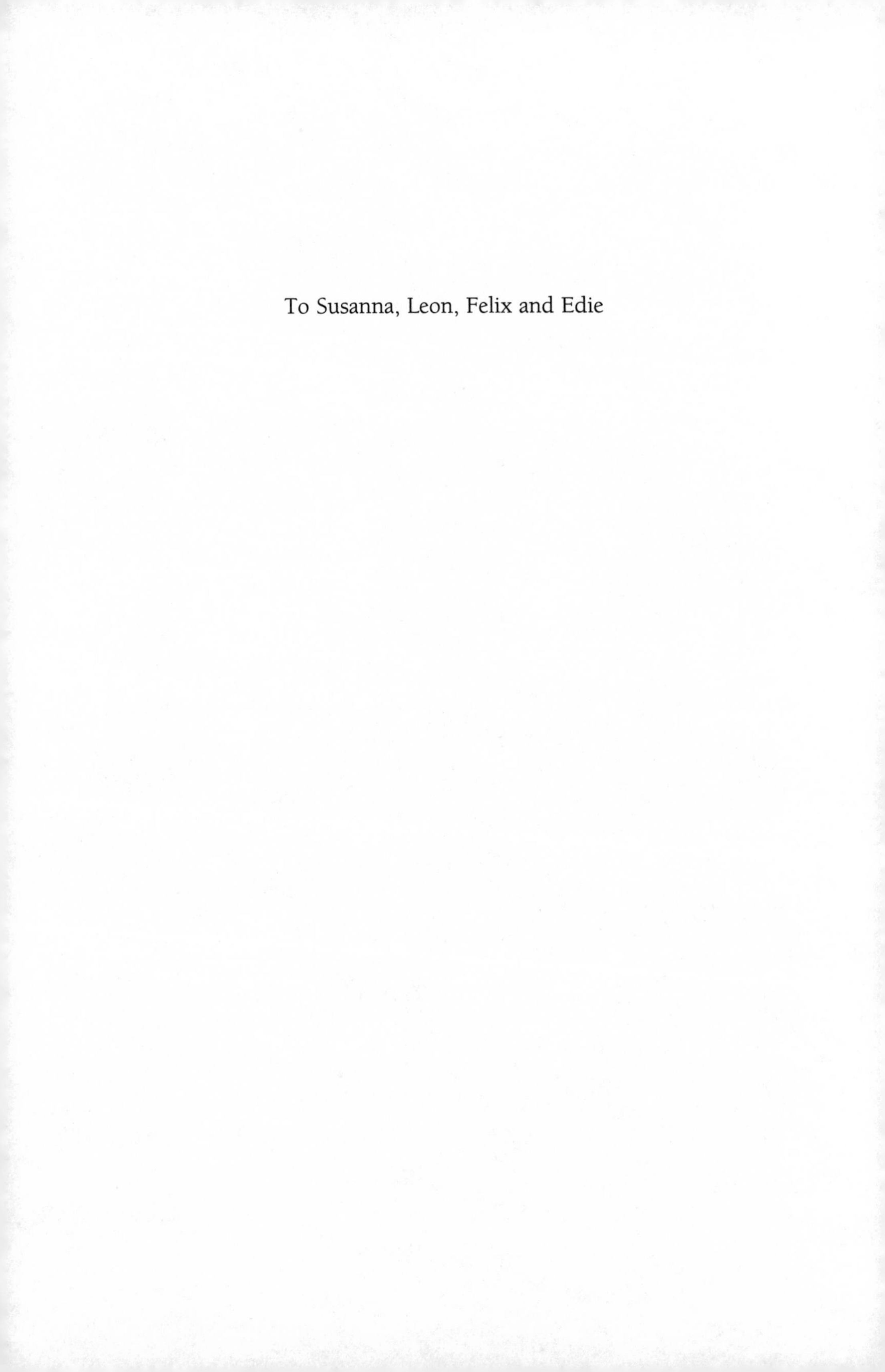

To Susanna, Leon, Felix and Edie

Contents

A science is any discipline in which the fool of this generation can go beyond the point reached by the genius of the last generation.

Max Gluckman

Foreword

On July 4, 2012, two large groups of physicists working at the Large Hadron Collider, the enormous machine near Geneva that smashes together two very energetic beams of protons in the hopes of creating matter never before observed on Earth, announced a momentous discovery. They had found the particle known as the Higgs boson.

On that day in 2012, the world that we particle physicists know and study changed forever. A prediction that Peter Higgs had made about fifty years earlier was confirmed, as was the theory of the mechanism that Higgs, the team of Robert Brout and François Englert, and several others had developed. This discovery helped physicists not only more fully understand the Standard Model of particle physics—the theory of matter's most basic elements and interactions—but it provided theoretical physicists like me with essential information about how to grapple with the physics that underlies the Standard Model in the hopes of advancing beyond the status quo.

But for experimenters like Jon Butterworth, the Higgs discovery changed the world in a more immediate fashion. The many physicists working on the two major general-purpose LHC experiments, ATLAS and CMS, had successfully completed their first major LHC goal: to find this particle, show it did not exist, or demonstrate that a more complicated or subtle model was at work. The discovery accomplished the first of these possibilities, but also meant that the experimenters who had been so successful in this first mission now had even more work cut out for them. They could now perform the detailed measurements that would determine the new particle's properties sufficiently well to either confirm theoretical predictions or determine that they did not precisely conform to expectations, paving the way for something new.

At the time of discovery, I was overwhelmed by what it meant for science—and also by the many questions I was being asked. As a response to this wonderful excitement and curiosity, I wrote a short book about the Higgs as a coda to *Knocking on Heaven's Door*, which I'd written while anticipating results from the LHC. *Higgs Discovery: The Power of Empty Space* was an opportunity for me to explain both the discovery and what it meant for theoretical physicists like myself, who predict and respond to experimental results in an attempt to piece together their implications.

But Jon Butterworth has a different story to tell. He is not a theorist but an experimenter who was actively working at CERN (the facility where the LHC is located) on the ATLAS experiment and who was privy to many of the internal discussions and activities that led to that thrilling moment in July when the results were announced. Jon is the ideal experimenter to tell the story of the anticipation and preparation, the team's experiences at the time of discovery, and the implications of the discovery for his colleagues and him. He is the rare scientist who can actively engage in research while also clearly explaining to the public what he is doing and why it is important. He participates in groundbreaking physics research, but he is also getting the word out through his blog and his writings for the *Guardian* newspaper in Britain and elsewhere.

In fact, I think I first heard Jon Butterworth's name from his public writing. Experimental collaborations at the LHC have a few thousand people, so theorists don't immediately know them all. He also shares my fondness for Twitter as a means of communicating scientific results, so we can learn some science from each other that way, too. We are not alone in this. I was amused when, while we were out for drinks, Jon introduced me to Mark, one of his colleagues, who then promptly informed me that I knew him already from his informative ATLAS tweets. He was right. Indeed without thinking I spouted off Mark Tibbetts' full name.

But one day in a conversation about analysis methods, I heard about some interesting analysis tools and learned that Jon was working on these, too, meaning he is not solely an excellent communicator but a true experimenter in the best sense of the word. Jon is someone who can talk to theorists and who also knows ATLAS inside and out. Most importantly for this book, he is someone who can successfully translate the process and the physics for the public while providing a sense of the experience of a physicist who is heavily involved in the cutting edge of the field.

In this book, Jon, with his delightfully nerdy self-referential humor that physicists—well, at least some of the better ones—often have, captures the wonder and elation that I and others experienced when first witnessing the machine and the experiments. Jon tells us what the LHC is, what it is designed to study, and why people work there. And his first-person account reveals what life was really like as a physicist at the LHC before, during, and after the discovery, from the initial circulating protons in 2008, the disaster that ensued nine days later that delayed the actual physics run for a year, the following year's hard work of fixing the machine and properly readying the experiments, and finally the completion of the actual physics run.

It is a great story, and Jon's telling of it not only gives readers a visceral feel for what it was like to be there as an experimenter participating in this enormous collaboration, but teaches a lot of physics along the way. Jon conveys the excitement, the anxiety, the sleep-deprivation, and the sense of satisfaction that went into the results. He describes how the LHC, twenty-five years into its history, was responsible for the discovery of an actual new particle in the universe—one that was predicted on purely theoretical grounds and found through the hard work of scientists and engineers.

Finding the Higgs boson was one of the most amazing experimental results of my lifetime. My colleagues and I still discuss over lunch the bizarreness of its actual existence. When first contemplated, it was a theory. The model could have taken many forms. The Higgs boson was part of the simplest versions of that theory—one that doesn't even seem to fully make sense when taken in the full context. Yet it was a precise model with specific predictions that could be searched for. In fact, by the time the LHC turned on—despite the theoretical misgivings—experimental results seemed to indicate that indeed the particle did exist and should be just barely accessible to the first major LHC implementation—even before the upgrade to higher energy that is now underway. And with the extended LHC run that finished in 2012, the anticipation culminated in the now-famous uncovering of the actual particle—buried in the mountains of data the experimenters had collected.

This book shares the joy of that discovery as well as the joy of science more generally. It also describes the challenges that science faces in the precarious political and economic climate of today. Jon's tales from the front lines of the debates over the role of science in Britain impart lessons that apply to all of us around the world. I hope Jon's book encourages people to value the amazing insights into nature that discoveries like the Higgs boson reveal, as well as inspires future generations to learn more about how our world really works.

Lisa Randall, Harvard University theoretical physicist and author of *Warped Passages*, *Higgs Discovery*, and *Knocking on Heaven's Door*

Introduction

There is a kebab restaurant in the Meyrin suburb of Geneva that has half a dozen pool tables. In early July 2012, I found myself playing pool with Tom Clarke, the science correspondent of Channel 4 News, one of the UK's major TV news bulletins, by way of trying to explain to him and his viewers the significance of the discovery we had just made at the Large Hadron Collider.

I still find that last sentence amazing – both the discovery and the huge public interest demonstrated by the fact that Tom, along with many other journalists, came out for a day and spoke to dozens of physicists. His report was the lead item on the 4 July bulletin.

The discovery we announced that day was a huge step forward in physics. The public interest was a significant milestone in people's increasing engagement with the science that lies behind our civilisation. I really mean the science, not just the technology but the processes of science – to what extent it is self-correcting, and what constitutes scientific certainty (very little!) and scientific knowledge (a lot!).

Meyrin is significant here because CERN, the European laboratory for particle physics, is just five minutes up the road. Meyrin village is quite picturesque, but the part Tom and I were in, Cité Meyrin, is a series of blocks of flats that would be a urine-smelling, graffiti-ridden concrete jungle pretty much anywhere else in the world. However, because this is Switzerland (just – by about 100m) it is a clean, orderly concrete jungle. It is also where many of the scientists working at CERN stay.

I work for University College London (UCL), but, along with many particle physicists from all over the world, I do most of my research at CERN. The UCL commuter flat is in Meyrin, and my colleagues and I spend a lot of time there. In particular, I ran a working group on the ATLAS experiment at CERN from October 2010 to October 2012, the period during which we got our first flood of high-energy data. During that time, I was there more or less every week.

This book is not a physics textbook; it is not a historical account of the discovery of the Higgs boson; it is not a diary; and it is not a manifesto for greater engagement between scientists and the general public. It does contain elements of all these things, however. You will learn a lot about particle physics and what it is like to be a particle physicist, about how science works (and occasionally doesn't), about how research sometimes struggles to thrive and survive, and about the people who do it, including a bunch of personal opinions from me. I hope it will also explain why Tom Clarke and much of the world's media descended on Meyrin that July.

To get that far, though, I need to introduce a number of interconnected and probably unfamiliar pieces of information. Some of them won't seem very relevant the first time they appear, like isolated pieces of a jigsaw puzzle, but as you collect them through the book, hopefully they will start to reinforce each other and in the end the full picture will emerge. And if I succeed, you'll have fun as you follow the story collecting the pieces – and gain a sense of excitement. Because fun and excitement are the two impressions that dominate my memory of the first high-energy run of the biggest scientific apparatus ever constructed: the Large Hadron Collider.

ONE

Before the Data

1.1 Why So Big?

The Large Hadron Collider (LHC) sits in a tunnel 27km (nearly 17 miles) long and about 100m (almost 330 feet) underground. If you know London, it might help you to know that 27km is about as long as the Circle Line on the Underground, and the tunnel itself is similar in size to the Northern Line. If that doesn't help, then try this.

Imagine setting off from Meyrin, on the Swiss–French border near the airport, and driving towards the French countryside. The Jura Mountains are in front of you, Geneva Airport is behind. As you pass the border, you also pass the main site of the CERN laboratory on your left, and if you look to the right you will see a big wooden globe that looks like a sort of eco-nuclear reactor (it's not, it's an exhibition space, though it is eco-friendly, apparently), and you might catch a glimpse of the building housing the control room of the ATLAS experiment. You will know it if you see it, because it has a huge mural of the ATLAS detector itself on the wall.

Big though it is, the mural is painted to only one-third scale. ATLAS is very large, and is hidden underground, positioned at one of the interaction points of the LHC. These are the points where the two highest-energy particle beams in the world are brought into head-on collision. ATLAS is one of the two big general-purpose particle detectors designed to measure the results of these collisions.

Continue driving. You may imagine yourself in a nerdy little white van with a CERN logo on the side if this helps.

Pass through the village of St-Genis and continue into the Pays de Gex, in the foothills of the Jura Mountains. You are now surrounded by the LHC. If you are imagining yourself in winter, you might see the lifts of Crozet, the little Monts Jura ski resort, chugging away ahead of you. (Mont Blanc is behind you on the horizon, but keep your eyes on the road.) Keep driving, bear right towards Gex, maybe pass through the villages of Pregnin, Véraz and Brétigny. After about 25 minutes' driving through the French countryside – longer if you get stuck behind a tractor – you will get to the village of Cessy, near Gex. Here you will find the top of the shaft that leads down to CMS, the other big general-purpose detector on the LHC ring. ATLAS and CMS are independent rivals, designed differently by different collaborations of physicists, but with the same goal: to measure as well as possible the particles produced when protons collide in the LHC. They were designed to cross-check each other's observations, and to compete head-to-head for the quickest and best results.

All this time, on your journey from ATLAS to CMS, you have been inside the circumference of the world's biggest physics experiment. You entered it at the border when you passed ATLAS, and have now crossed its diameter.

The LHC is designed to collide subatomic particles at the highest energies ever achieved in a particle accelerator. We do this to study the fabric of the universe at the smallest distances possible, which for reasons to be described later also implies the highest energies possible. Given that the experiment is designed to look at very small things, it might be a surprise that it is so big. Building a long tunnel is very expensive, so why not make a smaller one?

In fact, it is the length of the tunnel that limits the energy of the colliding beams. If you accept the fact that to study small stuff you need high energies (please do, for now at least), you can understand why the LHC needs to be so big just from an understanding of fairly everyday physics.

Particles travel in a straight line at a constant speed, unless a force acts on them. This is one of Newton's laws of motion. In everyday life it isn't completely obvious (Newton was quite clever to work it out), but once you are aware of it, it is easy to see it in action.

The reason it is not completely obvious in everyday experience is that on Earth practically everything that moves has forces due to friction and air resistance acting on it, and everything experiences gravity. This is why if you set a ball rolling, it will eventually stop. Friction and air resistance act on it to slow it down. And if you throw a ball in the air, gravity will slow it down and eventually drag it back.

But in situations where friction or gravity can be ignored, things are clearer. Driving a fast car, or even a nerdy CERN van, you clearly have to apply a force, via the brakes, to slow it down. And more relevantly in the context of the LHC, if you want to change direction, to turn a corner at speed, this can only be done if there is sufficient friction between the tyres of the van and the road. Otherwise, you skid.

The driver and passengers experience a rapid turn of a corner as a sort of 'pseudo-force'. The van is turning, but your body wants to carry on in a straight line, so you feel as though you are being pressed against the sides of the van. It would be more true to our understanding of physics to think of the sides of the van as pushing against you, to force you to change direction, pushing you round the corner along with the vehicle.

The combination of speed and direction is called velocity. And if you combine the velocity and the mass of the object (the van, for example, or the passenger), you get the momentum. The bigger the mass, or the velocity, the bigger the momentum, and if you want to change the momentum of something, you need to apply a force to it.

I am being deliberately vague about how the velocity and mass combine to give momentum. At speeds much lower than the speed of light, it is good enough to just multiply – momentum is mass times velocity – and this is probably the right answer if you are taking a school course in physics. However, the exact expression is a little different, and the difference gets more and more important as speeds approach the speed of light. Then you need Einstein and relativity (of which more later), rather than Newtonian mechanics. But don't try this in a van.

Regardless of that, the larger the desired change in momentum, the bigger the force has to be. Hence the brakes on a lorry need to be able to exert more force than the brakes on a van, because even if the velocity is the same, the mass of the lorry is bigger so the change in momentum involved in making it stop is bigger.

This is the situation of the protons in the LHC tunnel. These are the highest-energy, and highest-momentum, subatomic particles ever accelerated in a laboratory. Even though the mass of a proton is tiny, their speed is tremendously high. They are really, really determined to travel in a straight line. So, to make the two beams of protons bend around the LHC and come into collision requires a huge force. The force is provided by the most powerful bending magnets we could build.

Given this maximum force, there is then a trade-off between how sharp the bend in the accelerator is and how high the proton momentum can be. Back to the van: this is exactly equivalent to the fact that there is a maximum speed at which you can take a given corner without skidding. If the corner is sharp, the speed has to be low, but for a gentle curve you can go faster. This, then, is why the LHC is so big. A big ring has more gentle curvature than a small one, and so the protons can get to a higher momentum without 'skidding'. Or, in their case, 'catastrophically escaping the LHC and vaporising expensive pieces of magnet or detector'. Something to be avoided.

The maximum bending power of magnets is thus the reason that proton accelerators need to be large if they are to get to high energies. For the other commonly collided particle, the electron, there is another reason that is worth looking at.

Before the LHC was installed, another machine occupied the 27km tunnel under the Swiss–French border. This was LEP – the Large Electron–Positron Collider. (Positrons are the antiparticle of the electron, carrying positive charge, in contrast to the electron's negative. LEP collided electrons and positrons together. Incidentally, people occasionally accuse particle physicists of hyping-up their equipment, but these are very descriptive, even dull, names.) LEP was turned off in the year 2000 because it had explored most of the physics within its reach and could not increase its energy further. The reason it could not go higher was, as with all the protons, also connected to the size of the tunnel, but in a different way.

This is to do with the fact that electrons have a mass about 1800 times smaller than the proton. Now, at the highest energies that doesn't make any significant difference to the force required to bend them round a corner. This is because, whether they are electrons or protons, they are moving very close to the speed of light, so you need the full special relativity expression for momentum, and the net result is that the mass they have when they are at rest is irrelevant for calculating the required force. So that wasn't the problem.

The problem was synchrotron radiation. This is the energy radiated by charged particles when they are accelerated. It is a universal phenomenon, roughly analogous to the wave a speedboat makes when it turns in the water. As a charged particle accelerates round a corner, photons fly off and carry away energy.

The effect is actually much more pronounced for particles with low mass. The amount of synchrotron radiation given off when a particle accelerates depends very strongly on the mass: if the particle mass drops, the energy loss increases by the mass-drop to the fourth power. So, as the proton mass is 1800 times bigger, the energy lost on the bends for electrons is (1800 x 1800 x 1800 x 1800) or about 11 trillion times larger than it is for protons.

As the electrons and positrons squealed round the corners of LEP, photons were radiated this way, and with every revolution of the beam around the ring, more energy had to be pumped in to compensate. This is done by radio-frequency electromagnetic waves confined in big metal structures at intervals around the ring. Electric and magnetic fields oscillate in these structures precisely in time with the passing of the bunches of electrons, so that every time a bunch arrives it gets a kick from the field. This is true in all such machines. But at some point you reach a beam energy where so much is lost in synchrotron radiation that the electromagnetic waves in those structures cannot replace it. That's your maximum collision energy. LEP hit that wall.

This is where the size of the tunnel comes in again, of course. A 27km tunnel has a rather gentle curve. If it were smaller, the bends would be sharper, the acceleration would need to be bigger, so the energy lost through synchrotron radiation would be greater, and the maximum collision energy would be lower.

As an aside, this synchrotron radiation is very useful in other contexts. The Diamond Light Source at Harwell in Oxfordshire, in South East England, for example, was built to produce it intentionally. The radiated beams of photons are used to study atoms, crystals, molecules, materials and surfaces. Many machines and laboratories originally built to study particle physics have been converted to become light sources once they have been superseded in the quest for higher energies. I have reason to be grateful for this personally, in fact. I did my doctoral work in Hamburg, at the DESY (Deutsches Elektronen-Synchrotron) laboratory. The particle physics of interest there at the time was the HERA electron–proton collider, where I worked in the ZEUS collaboration, the team of physicists responsible for one of the main particle detectors at the laboratory. But my then girlfriend was a crystallographer, using synchrotron light to work out the structure of proteins and other stuff. Because of the symbiotic relationship between particle-physics accelerators and synchrotron light sources, there is a branch of the European Molecular Biology Lab at DESY, and after a high-level discussion in the crowd at a St Pauli football match, Susanna managed to get her PhD supervisor to send her to Hamburg for most of her research. We've been married 20 years now, and it's all very fine and romantic. But synchrotron radiation is still a pain in the arse if you want a high-energy electron beam.

So, LEP was shut down in 2000 and dismantled, and installation of the LHC began. The LHC can get to higher energies because it collides protons with 11 trillion times less of a synchrotron radiation problem, but it requires the most powerful bending magnets you can make if you want to get to the highest possible momentum.

The formal approval for construction of ATLAS and CMS was given on 1 July 1997 by the then Director General of CERN, Chris Llewellyn Smith.[1]

LEP had been good, but the protons promised more.

---

[1] Incidentally, a man who previously, as head of physics at Oxford and afterwards as provost of UCL, seems to have had a period of following me around and being my boss.

Glossary: The Standard Model Particles and Forces

If you just want to crack on with the story and don't mind the odd unfamiliar word, you can skip these 'Glossary' bits. But without knowing something about the Standard Model, some of it might not make much sense.

The Standard Model of particle physics is our current best answer to the question 'What is stuff made of, if you break it down into its smallest components?'

Start with anything – a rock, the air, this book, your head – and pull it into its component parts (I recommend this remains a thought experiment). You will find fascinating layers of structure, micro- and nano-scale bits and pieces: fibres, cells, mitochondria.

You will eventually find molecules. With enough energy you can break them apart into component atoms. Atoms consist of a dense nucleus surrounded by electrons.

With a bit more energy, you can separate the electrons from the nucleus. With more energy still, the nucleus can be broken into protons and neutrons. With still more energy (and now you do need a big collider!), you can see quarks inside those protons and neutrons.

We have never managed to see anything inside a quark, or break one into pieces.

If, at the 'atom-smashing' stage, we had ignored the nucleus and tried breaking up the electron, we'd have reached that point earlier. We have never managed to see anything inside an electron, or break one into pieces. This – the fact that we haven't managed to break one yet – is our working definition of what it means for a particle to be 'fundamental'.

And a key point is that wherever we had started, with whatever material, we would have ended up with electrons and quarks. In the Standard Model, they are the stuff that everything is made of, and they themselves are not made of other stuff.

You will come across a lot of particles in this book, but remember, there aren't many different kinds of fundamental ones when you get right down to it.

Electrons are an example of a class of particles called leptons. There are also muons and taus, which are just like electrons only heavier. The only other leptons are the three kinds of neutrino. Neutrinos do not interact much with other matter, but there are lots of them around. More than a trillion neutrinos pass through you from the Sun every second.

The other class of fundamental-matter particles consists of the quarks. There are six of them, too, just as there are six lepton types. They are called up, down, strange, charm, bottom and top, becoming more massive as you go (but peaking on whimsy in the middle).

Protons and neutrons are made of up and down quarks. Quarks are never found out on their own, they are always stuck together in bigger particles. These particles, the ones made of quarks, are generically called hadrons (hence the Large Hadron Collider, which mostly collides protons but occasionally collides atomic nuclei, which also have neutrons inside).

Those are all the matter particles we know of. They all have antiparticle partners, and they all interact with each other – attracting, repelling, scattering – via forces, which are carried by another kind of particle – vector bosons.

The electromagnetic force is carried by photons (quanta of light) and is experienced by all charged particles. That is, everything except the neutrinos.

The strong force, which holds protons, neutrons, and atomic nuclei together, is carried by gluons, and is only experienced by the quarks.

The weak force is carried by W and Z bosons, and all particles experience this. The weak force is responsible for radioactive beta decay, amongst other things, but because it is weak, it does not feature much in everyday life. Even so, it is crucial to how the Sun works.

To make the Standard Model work, and in particular to allow the fundamental particles to have mass, another unique and completely new object is also required – a Higgs boson. The hunt for this is, of course, the main topic of this story and I'll say much more about it later.

Gravity doesn't fit into the Standard Model. It is described by Einstein's theory of general relativity, but we do not know how to make a working quantum theory of that.

Those are the actors on the stage of the universe. There are lots of open questions in physics, but an astonishingly wide array of data – most of physics, chemistry and biology – from very large to very small distance scales, can be described astonishingly accurately by just these elements: quarks, leptons and the four forces between them, and the Higgs boson.

1.2 The 'No Lose' Theorem

I didn't begin working seriously on the LHC until around 2001. That was about nine years before we got our first high-energy collisions. Believe it or not, this makes me a bit of a Johnny-come-lately to the experiment. Options for a large hadron collider had already been considered in the design of the 27km tunnel for LEP, and were mentioned in the LEP design report in 1984, when I was just finishing secondary school and moving to sixth-form college. There would be many years of scientific, technical, financial and political discussions, followed by R & D, simulation and more politics, before the LHC gained approval in 1997.

Back in 1997, I had just moved to London from Hamburg and was still completely absorbed in work at HERA. It is a feature of big collaborations that you accumulate responsibilities along with a bank of experience and expert knowledge that can make it hard to disengage. It is difficult to climb a new learning curve on another experiment, with its confusing software and hardware and unfamiliar physics. Sometimes you need a bit of a shove to really start doing something else.

For me, bizarrely, the shove was the birth of my first child. This was such an overriding priority that I managed to say no to a whole bunch of managerial and technical roles within the ZEUS collaboration. I wanted no responsibilities that would conflict with the terrifying challenges of looking after Susanna during her pregnancy and of being a dad after it.

As it turned out, the whole thing went very smoothly and was generally wonderful. So, as a bonus, I had lots of free time to think creatively about physics. One of the things I'd long been wanting to do was read enough and think enough to get my head around physics at the LHC, which was by then under construction at CERN. This holiday from HERA heat was the opportunity. With a couple of friends, Jeff and Brian, also HERA physicists,[2] I'd been thinking about what we might do – what the most exciting things to study would be. We were all very sceptical about new 'Beyond the Standard Model' physics and were keen to work on measurements of real things that would actually happen, rather than seeking evidence for speculative ideas to which we accorded little credibility. I think this may have been because we all came from a HERA background, where precision measurement was the main goal. Although to be honest, the main legacy of LEP was also precision measurement, so maybe it made no difference.

Anyhow. Not only did we not believe in such things as supersymmetry, or large extra dimensions, or Technicolor, all of them speculative extensions of the Standard Model designed to solve some of the problems with it. We didn't even believe in the Higgs boson – an integral part of the Standard Model, but one lacking experimental verification. So we asked ourselves, 'What is the most important and interesting thing to measure if there are no new particles?' A pessimist's approach, perhaps, but still fun.

The answer we came up with[3] was vector-boson scattering. This is a peculiar and rare scattering process that is expected to happen occasionally in very high-energy collisions, and it lay behind what was called the 'no lose' theorem at the LHC. It is very deeply connected to the reasons why the Higgs boson is so important. So it wasn't a bad choice for a

---

[2] Jeff Forshaw and Brian Cox, who amongst other things have also written physics books, though not about this yet.

[3] After some reading around – I'm not claiming we were the first to think of this!

first bit of LHC physics for us to look at, and it's worth spending a bit of time on now.

Vector bosons are force carriers. The photon, which is the quantum of light and carries the electromagnetic force, is a vector boson. Of more interest here, though, are the W, and to some extent the Z, bosons. These carry the weak force, and one of the oddest things about them is that, unlike the photon, they have mass.

In a proton–proton collision at the LHC, you have to picture two quarks, one from a proton in each beam, zooming towards each other. There is a small but non-zero chance that, as they do this, each will radiate a W boson. There's an even smaller, but still non-zero, chance that these W bosons will hit each other. That is vector-boson scattering – WW scattering in this case. It could happen with Zs or photons too. There are a bunch of different ways the bosons can bounce off each other, or fuse together and break up again. As is always the case in quantum mechanics, all the possibilities have to be taken into account and combined[4] – sometimes they add up, sometimes they subtract from each other. Put the whole thing together and you get the probability of the WW scattering occurring.

The 'no lose' theorem came from this calculation. Some of those scattering possibilities include a Higgs boson, and at the time there was no direct evidence for such a beast. However, if you do the sum and do not include a Higgs boson, then as you go to higher and higher energies, the probability of WW scattering grows and grows.[5] At some point you get nonsense answers involving probabilities bigger than one, or infinities. That is just a sign that your theory is broken – there won't be infinities in nature – but what it meant was that either a Higgs boson would be discovered at the LHC, or some other new physics would come into play and keep the calculation sane.

---

[4] This includes taking into account time orderings other than the quarks-emit-Ws-that-then-collide one I gave here.

[5] The possibilities involving the Higgs would contribute with a negative sign, and so they would stop this happening.

So in the pessimist's scenario of no Higgs boson, no black holes, whatever, measuring WW scattering might well be the only, or best, clue as to what was going on. It certainly had to involve either a Higgs boson or something else new – hence the 'no lose' theorem. By studying these scatterings we were certain we would discover some interesting physics.

Measuring WW scattering properly would be difficult, and we found lots of fun challenges. The Universtiy of Manchester had done a deal with Apple and had a big new farm of Macs that had just started running Unix (OSX), making them useful to us (if still more expensive than the Linux boxes everyone else had). I have fond memories of sitting in a flat in Saddleworth, feeding the farm with lots of simulation jobs to test our ideas, then popping over the road to the pub for beer and dinner to argue about newer ideas. This was still before my son was born, but I had already divested myself of lots of other responsibilities. We submitted the paper in January 2002 and it was more or less ignored for six years, though it later became more fashionable and I'm very proud of it. One of the ideas we used turned out to be quite widely useful and would feature in the Higgs search itself.

1.3 People Are Going to Be Interested in This . . .

While work was going on at CERN and around the world to construct the LHC and its detectors, it became increasingly obvious that quite a lot of people were going to be interested in the project, for all kinds of reasons. For the engineering and science, of course. But also because of the sheer scale, including the cost of the thing. The international collaboration and the sociology of several thousand physicists working together were intriguing to quite a few people, including academics in social sciences. Plus, of course, there were the two or three delusional publicity-seekers who thought, or claimed to think, that we were about to destroy Switzerland. Or the world. Or the entire universe.

The last bunch – the delusionists and conspiracy theorists – were bound to get lots of media coverage, 'because of balance'.[6] The only way to deal with that is to get real information out there. Also, since the European taxpayer had been investing something like the equivalent of a billion euros in CERN every year, we really owed it to people to explain what we'd done with the cash, and why.

Thoughts like these were passing through the minds of many people involved, including, I am sure, James Gillies, the head of communications at CERN, and many good science journalists. This is presumably why, in 2008, the doors of CERN were flung open to the world's media for 'Big Bang Day', when we switched on the machine.

Such thinking was certainly one reason why I agreed to be part of a series of short documentary films called *Colliding Particles*. These were a sort of 'fly on the wall' affair that started in the summer of 2008. Mike Paterson was the cameraman, producer, interviewer and director – everything, in fact, except animator and occasionally soundman. He won some support from the Science and Technology Facilities Council (STFC), the research council that funds particle physics in the UK, to make the films. They were aimed at schools, specifically at a then new part of the curriculum based on learning how science works. Apparently pupils would learn this by watching some physicists from behind Mike's camera.

Actually, it worked out very well. Amongst other things, Mike has some sort of genius for cutting and matching a long ramble by me to amazing pictures that make me appear coherent. There is roughly five minutes in the first film where I speak without ever finishing a sentence properly, while shots of the LHC and of the ATLAS detector being assembled fill the screen, imparting to the viewer a sense of the vision and wisdom I am pouring forth. Well, that's how it looked to me and my mum, anyway. The films also featured Adam Davison, who at the start was my PhD student, later a postdoc, and Gavin Salam, a theorist in Paris,

---

[6] To quote David Shiffman: 'World's leading experts say there's a problem with false balance in environmental journalism, but Steve disagrees.'

and were (at least to start with) loosely based around a paper we'd written in 2007 with Gavin's student, Mathieu Rubin.

I mention all this now to show that, despite strong reservations from some particle physicists, we were taking public engagement quite seriously. The film was one of several initiatives at the time, a level of activity unprecedented in particle physics. Mike, Adam, Gavin and the films will pop up from time to time as we progress towards discovery.

Anyway, here we were, after many years of R & D and eight years of construction, at the switch-on.

It was 10 September 2008. In CERN, the control room was packed with journalists and Brian Cox. BBC Radio 4, the UK's major news and current affairs channel, made a day out of it – 'Big Bang Day'. I had come back from CERN and was in Westminster, in a big hall with the minister (John Denham, the UK's Secretary of State for Innovation, Universities and Skills), many other of the great and occasionally good, and more journalists. This was all very exciting and totally new territory for us in terms of engagement with the media and politicians, but still the most terrifying and exciting thing was the fact that after so many years of preparation we were finally about to switch on our experiment.

The switch-on was pretty closely choreographed, with Lyn Evans, the LHC project leader, the ringmaster in the LHC control room. The beams were to be sent into the LHC octant by octant. That is to say, initially the beams would go an eighth of the way around the 27km tunnel and hit a beam blocker. Then a quarter, three-eighths, and so on until hopefully they got all the way around and came back again, registering two dots (one on the way in, one on the way back) on a scintillation counter that was the centre of attention for thousands of physicists and a significant chunk of the world's media.

One nice thing was that when the beam hit a blocker in front of one of the detectors, a spray of particles would be produced that would register in the detector – the first beam activity we'd seen in these highly complex and sensitive pieces of kit. Less nice from my point of view was that because Lyn chose to send the clockwise beam around first, ATLAS was the last in line, and therefore the last to see particles. Still, the next step after us was a full circuit. Lyn counted down: 3 . . . 2 . . . 1. A moment of nerves when nothing appeared . . . then bingo! Two dots. The most exciting two dots I've seen before or since. For the first time, a beam had successfully completed a circuit of the LHC.

As the day progressed, beams were sent round in both directions and stored successfully. Sheer exhaustion at Westminster led us to the pub next door, though the accelerator teams at CERN were still hard at work. I will never forget drinking a beer at lunchtime and seeing progress updates of my own physics experiment on the BBC news ticker at the bottom of the pub TV screen. It was hard not to be triumphalist about it all. The headlines the day after, declaring that we hadn't destroyed the world, were fun if a bit premature (we hadn't yet collided the beams, after all), but we had got ourselves a working experiment and we had managed to share our excitement with the people who were paying partners in the enterprise. I was looking forward to welcoming the new PhD students to UCL with the promise of imminent data.

Nine days later, it had all gone catastrophically wrong.

1.4 Breakdown

As I have already described, the limiting factor on the maximum energy the protons can get to in the LHC is the centripetal force – the force needed to bend them round the corners so that they stay in the ring. This force is provided by huge magnets. Imagine whirling a brick round your head on a thin piece of string. If you whirl it too fast, the string will break. The protons are the brick, our magnets are the string. We really do not want them to break.

In days after the start-up, many tests were performed on the LHC. In particular, when the beams were circulated on 10 September, the magnets had not been running at full strength.

The magnets are electromagnets, meaning that the magnetic field is generated by circulating electric currents. As well as the impressive engineering and industrial know-how that goes into making them (there are 1232 dipole magnets to bend the beam, each of them is 15m long and has a mass of 35 tonnes), there is a lot of physics here.

The facts that electric currents create magnetic fields and that magnetic fields bend electric currents were observed and measured by Michael Faraday in the 19th century, and built into the theory of electromagnetism by James Clerk Maxwell. Maxwell's equations are one of the highlights of a physics degree, and are arguably the first expression of a mathematical unification of two apparently different forces (electrostatics and magnetism), setting a trend that physics has followed ever since.

To produce the force required to bend the proton beams at the LHC, very high currents are needed. At full power the magnets need to carry a current of nearly 12,000 amps (A). This is about half a million times bigger than the current drawn by a typical incandescent household light bulb.

When an electric current flows through a normal material (such as the light-bulb filament), the electrons carrying the current collide with the vibrating atoms of the material. This makes the electrons lose energy and the atoms vibrate more vigorously, heating up the material. This is electrical resistance, and it is a big problem if you want a current of 12,000 A. Essentially, any normal material will vaporise.

The discovery of superconductivity changed this. Superconducting materials offer zero resistance to the flow of an electric current. This is a rather amazing quantum mechanical effect, understood in the Bardeen, Cooper and Schrieffer theory. At low temperatures, the electrons form pairs and start to behave like bosons.[7] This allows them all to sink into the same quantum state – a condensate – and overlap with each other. At that point, it takes quite a lot of energy to have any effect on a pair, because that would mean changing the whole condensate, which is in a coherent quantum state. In general, the collisions with the material don't have enough energy to do this – at least not when the material is very

---

[7] See Glossary: Bosons and Fermions (pp.31–3).

cold, when the atoms in the material are hardly vibrating at all. So the pairs of electrons flow on unimpeded, losing no energy and experiencing no resistance.

The magnets in the LHC are superconducting. They are cooled to a temperature of 1.9 kelvins (-271.3°C), using pressurised liquid helium.[8]

On 19 September, the LHC team were testing the magnets up to full electric current – the current at which they could bend beams around the LHC at full energy. The LHC operates as eight independent sections, which can be powered, warmed or cooled separately. They had commissioned seven of these octants to full current and were on the eighth and last, almost ready for first collisions.

At this point, all information from the monitors and sensors in that sector suddenly ceased.

Watching the dashboard remotely in London, I just saw a note that 'first collisions' had been delayed for at least a few days. But the reality was much worse. There had been a catastrophic explosion. Several of the huge magnets had been ripped out of their concrete moorings. We were not going to get our first proton–proton collisions for more than a year.

I had to explain to various journalists, including a live phone call to breakfast TV, what had happened. To tell the truth, other than that it was very bad, we didn't completely know at that stage. By far the worst experience was having to stand up in front of the PhD students and tell them they wouldn't be getting any collision data for a while longer. I phrased it as 'two steps forward, one step back', but the step backwards felt enormous.

The full story gradually emerged days and weeks later. There had been a fault in one of the connectors between two magnets. This was due to a flaw in the welding. The connector developed a small electrical

---

[8] This is, as is frequently pointed out, colder than outer space. The cosmic microwave background has a temperature of 2.7 kelvins. However, it is not, as is sometimes said, the coldest place in the universe. Apparently the Boomerang Nebula is at 1 kelvin. I have no idea why.

resistance. On its own that would have been a serious but not catastrophic problem. Part of a superconducting system suddenly developing a resistance is an occupational hazard of the technology; it leads to what is called a 'quench', and the superconducting magnets have elaborate protection against quenches, meaning that the enormous energy in the electric currents is safely dissipated before it heats up and damages the magnet.

Unfortunately, the quench protection didn't extend to the connectors. The current was not safely dissipated, and the connector was vaporised.

Again, on its own this would have been a big problem, leading to months of delay, but not a catastrophic one. However, the huge current now had nowhere to go and sparked across the gap left by the vaporised connector. The spark punctured the liquid-helium containment vessel and suddenly tonnes of pressurised liquid helium became a gas. Very quickly. This was the explosion, powerful enough to tear some of those 35 tonne magnets from their concrete moorings. Several of them were destroyed or damaged, and the precision instrumentation and delicate cryogenics were turned into a mess of twisted metal. It all meant that we had a very long wait, and a lot of work to do, before the machine would deliver physics results.

1.5. While We Were Waiting . . .

The aftermath of this disaster was quite instructive.

As with any big project, there are people who object to the amount of resources spent. There are people who resent the huge public and media interest that is denied to lots of other good science. Also, apparently, particle physicists are sometimes perceived as arrogant, though I can't think why. For whatever reason, while there was much genuine sympathy for us disappointed LHC physicists, there was also *Schadenfreude*.

And remember, a substantial section of the particle-physics community thought the whole media- and public-engagement exercise around the start-up a big mistake. Many colleagues saw it as at best a hostage to fortune, and at worst unscientific media hype. The failure of the machine nine days after such a high-profile public event must have seemed a massive opportunity to say 'I told you so'. I was mainly just miserable about the delay, and did indeed have moments when I felt we'd made massive idiots of ourselves in a very public fashion and should have kept quiet until we had the results.

In October 2008, very soon after the catastrophe, we had the 'inauguration' of the LHC. This was a very odd event. It was a formal celebration, planned before the disaster. It was held in the huge magnet-testing hall, soon to be reopened to test (re-test) the repaired and refurbished magnets, which would be needed to repair the wounded machine below our feet. Despite being very interested in CERN and the LHC, Lord Drayson, the UK science minister, didn't show up. To be honest, I don't blame him. Frankly, it was a depressing affair.

But these feelings gradually changed.

The embarrassment became almost a source of pride. The failures in the magnet connector should not have happened, of course. But listening to accelerator physicists and engineers diagnose and discuss the systems involved just emphasised the amazing complexity of the LHC and the amount of new technology that had been integrated, on an industrial scale, into this huge machine. Not only were we on the edge of physics, we were on the edge of engineering, too. Also, no one had been hurt. In fact the whole LEP and LHC civil-engineering project, which was on the scale of the Eurotunnel construction, was carried out with remarkably few casualties.

Research carries risks. As the project leader Lyn Evans said, many times in many interviews, the LHC was its own prototype. Nothing like it had been done before. A notable feature of the wide-eyed physicists staring into multiple cameras on 10 September was that we were more nervous and excited about the possibility our experiment might, or might not, work than we were about our live-TV debut. The nearest equivalent is the visible nervousness on the faces of space scientists when a precisely engineered satellite they have built is making its way into orbit on top of a plume of flame.[9]

This was really doing science in public, and the reality is that science is not a seamless progression of triumphant advances. Two steps forward, one step back, indeed.

From the point of view of the media, this twist just prolonged a good story, and they were remarkably reasonable in the way they treated us, despite (or including, really) a certain amount of mocking on the occasional game show.

As for the misery . . . well that was harder to shake, but there were some important things to be getting on with. We'd locked ourselves into an obsolete version of the Linux operating system, fearing to upgrade with data imminent. We could fix that – and a lot of other bugs frozen into our software because we didn't have time to fix them properly. One of the most significant things we did was to get rid of an obsolete jet finder. But I realise that's going to take some explaining.

Glossary: Quarks, Gluons and Jets

Jets are what quarks and gluons make when they try to escape.

Every proton (in fact, every hadron) is made up of quarks stuck together by gluons. As already mentioned, these gluons are the force carriers of the strong force, just as photons carry the electromagnetic force, and the W and Z bosons carry the weak force. The role played by the electric charge in electromagnetism is played by a quantity called 'color' in the case of the strong force. This color has nothing to do with the colour we perceive with our eyes, and I'll use the

[9] In a display cabinet at UCL's Mullard Space Science Laboratory, they have the twisted remnants of delicate electronics from the first version of the Cluster mission, fished out of the Kourou swamps in French Guiana after the Ariane launch failed in 1996. Chastening stuff. But Cluster flew again, as Cluster II, and did the science it was built for.

US spelling to distinguish between them, since the physicist who introduced it, Oscar W. Greenberg, is an American. However, there is an analogy in there.

For electric charges, there is only one way to get a neutral charge – you have exactly as many anti-charges (negative charges) as positive charges. So an atom is neutral because the number of protons in the nucleus (carrying one positive charge each) is exactly equal to the number of electrons in the cloud around the nucleus (each carrying a negative charge). The result of adding up all the positive and negative charges is zero – they cancel each other out, and the atom is neutral.

For color, the same thing can happen. If we (arbitrarily) name one of the colors 'red', then there is also an 'anti-red' color (which you might want to call cyan since it's the complementary colour, though personally I think that pushes the analogy between the quanta of charge for the strong force and visible colours a bit too far). Color-neutral objects called mesons can be made by combining a color with its anti-color.[10] But one difference between electromagnetism (more correctly known in its quantum version as quantum electrodynamics, QED) and the strong force (known as quantum chromodynamics, QCD) is that there is an additional possible way of getting a color-neutral object.

There are three possible colors, often (still arbitrarily) called red, green and blue. If you have one of each, you also get a color-neutral object. This is where the analogy with mixing three primary colours to get white comes in. Protons and neutrons are made this way. They contain three quarks, one of each color, and so also end up being colorless. Particles made of three quarks like this are called baryons

---

[10] The name meson comes from the Greek *meso*, meaning 'medium', because its mass is lighter than that of protons and neutrons but heavier than an electron's. A meson contains a quark and an antiquark, and it is color-neutral because these will have, for example, red and anti-red color respectively, which cancel each other out. Or, if you prefer, red and cyan, making white.

(from the Greek, meaning 'heavy') and protons and neutrons are the most common examples. Mesons and baryons are both subclasses of hadrons, of course, since anything made of quarks is a hadron.

An odd feature of all this is that quarks are never observed on their own, a long way away from other quarks. They are always confined inside color-neutral hadrons of one kind or another, due to another peculiarity of the strong force.

Most of the fundamental forces get weaker with distance – the attractive force between a positive and a negative electric charge, for example, is weaker the further away from each other the charges are (it falls off as the inverse of the separation squared – $1/r^2$). But the strong nuclear force is different. The force between two quarks actually gets stronger as you pull them apart. It is as though they are attached to each other by an elastic band or a piece of string. As they move apart, the string becomes taut, and a large amount of energy is stored in the string tension.

Inside the LHC, when two quarks inside protons bounce off each other, they head away from each other at practically the speed of light and with an enormous amount of energy. Initially, the 'string' is slack and they feel very little force. This phenomenon is called 'asymptotic freedom', and the 2004 Nobel Prize in Physics was awarded jointly to David J. Gross, H. David Politzer and Frank Wilczek 'for the discovery of asymptotic freedom in the theory of the strong interaction'. It means that when quarks are inside the proton, you can for some purposes and to some approximation treat them as though they are free, as though they aren't bound together at all.

However, that apparent freedom ends rather quickly when you try to remove a quark from a proton – for instance by hitting it with another quark from another proton going the opposite way around the LHC. Though they fly away from each other initially (and even radiate more gluons and quarks as they accelerate), the string gets taut almost immediately, and the quarks and gluons know they aren't really free.

What happens next is intriguing, though. There is so much energy stored in the tension of the string between two quarks, as the force pulling them back together doesn't fall off with distance, that it becomes energetically possible, and indeed favourable, to make a new quark and an antiquark. The cost of doing this in terms of energy (E) is the mass of the quark plus the mass of the antiquark, multiplied by the speed of light squared ($E = mc^2$, but you probably knew that). But the benefit is that you can have much shorter strings and so much less potential energy stored in the string tension.

You can think of the quarks as being the ends of the string. They fly away from each other until at some point the string snaps and two new ends (new quarks) are produced.

Eventually, we see a spray of hadrons. You might think that this is a bit useless if we really want to see what is going on with the fundamental particles – the quarks and the gluons and so on. But all is not lost. Because the initial quarks get kicked so hard, the sprays of hadrons are shaped into narrow jets. All the splitting and production of new quarks and gluons shuffles energy around, but the amounts of energy shuffled that way are much smaller than the initial kick the quarks get from the collision. So in the end, the direction of the jet reflects pretty closely the initial direction of the quark.

Of course, 'pretty closely' is not a very scientific term. We need to quantify that and be as precise as possible about it. Jet algorithms are the tools that let us do that. They give a recipe of how to combine the observed hadrons produced in a collision, to get objects (jets) with energy and momenta that can be compared to a theoretical prediction. You can imagine many ways of doing that, but some ways are definitely better than others.

One issue in designing (or choosing) a good jet algorithm is the fact that the theory really doesn't know how to predict what happens at low energies. These low energies correspond to (relatively) long distances, where the strings are snapping, hadrons are being formed and lots of low-energy gluons get thrown around. Since you can't predict how many of these low-energy gluons might be produced – and since anyway that is not something we can measure, or want to measure, and it will fluctuate a lot – having a jet finder that is insensitive to the number of low-energy gluons seems like a good idea. In fact it is essential. The jargon for this insensitivity is 'infrared safety'. One of the things we did while the LHC was under repair was switch to an infrared-safe jet algorithm.

1.6 Names, Inertia and the Media

In 2008, the main jet algorithm used in ATLAS (and CMS) analysis code was not infrared safe. After the breakdown, we managed to change this to a newer, better, infrared-safe algorithm. This would make a huge difference to the quality of the physics we could do later.

You might wonder why, if it made such a big difference, we didn't change earlier. It's an interesting question, and the answer tells you something about doing science in very large collaborations (as well as something about physics).

Back when I was still in Hamburg working on HERA, LEP was still running at CERN and the Tevatron in Chicago was finding the top quark, proto-collaborations had formed to design and propose possible detectors for the coming LHC. ATLAS was formed out of two of these – EAGLE and ASCOT. This kind of thing often happens. It is very important not to use up your best collaboration name on the first proposal, since you will almost certainly have to merge with some other proposal and therefore have to pick a new name at some point. I can only presume CMS made this mistake. Probably their proto-collaborations were called cool things like TITAN or JOR-EL[11] but they had to merge so often they ran out of ideas and ended up as CMS.

Anyway, ATLAS is a good name, so well done, Peter Jenni et al.

---

[11] They weren't.

Marginally more important than choosing the name is showing that the detector will be able to do the physics required. Will it have a good enough resolution? Will there be enough bandwidth to read out the data? Will there be enough detectors in the right places to measure everything you want to see? And will you be able to afford it? To answer these questions convincingly, you have to write a technical design report. To do this and persuade people that you should be allowed actually to build your ideas, you need lots of results from test beams, where you fire particles into prototypes to show you understand them. But you also need huge amounts of software, some of it to simulate what the physics and the detector might look like, and some of it to reconstruct from the simulated data (or from test-beam data) what the measurement might eventually look like.

This means that by the time first data seem imminent, there are people who have been working on the experiment for ten or more years, and they have grown used to using the tools that were available ten years earlier. Attempts to change this will be met with huge inertia, even when there are much better tools available and when physics has moved on a lot due to data from elsewhere during those ten years. When the priority is to get everything ready for first data, the reluctance to change or reinvent anything is understandable. This is the state we were in on ATLAS in 2007.

The jet algorithm is just such a tool. The understanding of jets and the strong interaction, QCD, improved greatly in the 1990s and 2000s, due to a lot of work by theorists and a lot of data, mainly from HERA and LEP. The problems with infrared safety were understood and a new generation of jet finders was proposed. Unfortunately, ATLAS (and also, to a large extent, the Tevatron experiments) had already started using the old jet algorithms. And the new algorithms had problems: some of them were too slow, and lots of them made jets with irregular shapes, which made understanding the experimental resolutions and efficiencies harder. The HERA, LEP and Tevatron experiments did make measurements with the new algorithms (in fact HERA and LEP eventually moved over to them completely), but there were doubts about whether they could really be used at the LHC, and these doubts, combined with both inertia and the pressure of time, meant we did not change in time for 2008.

Of course, everyone knew that once we'd taken some data with the old algorithms it would only get more difficult to change to the new ones. So once we had an unlooked-for extra year before data arrived, it was a golden opportunity to switch, once and for all, to the new, better technology. Crucial, too, was the fact that the problems with speed and irregular-shaped jets seemed to have been solved.[12] Coordinated by our jet conveners, dozens of postdocs and students began checking whether the new algorithm really worked, not just theoretically but in the software ATLAS would use to select and analyse the data. Some of those people were working at UCL, and I edited the enormous internal note that, taken as a whole, finally persuaded everyone to make the switch. I think this was time very well spent. It certainly cheered me up a lot.

For me, another galvanising event, not directly related to the LHC, was a meeting on 18 May of the Skeptics in the Pub at Penderel's Oak in Holborn, London, just down the road from UCL. This was the first Skeptics in the Pub meeting I'd been to. It's a bit of a random thing to throw into a more-or-less chronological run-through of the LHC story, but it does connect, trust me.

Simon Singh, a well-known science writer with a PhD in particle physics – he worked on LEP at Cambridge with some of my present ATLAS colleagues – was being sued by the British Chiropractic Association (BCA) for writing that it was happily promoting what he felt were 'bogus treatments' without taking account of the lack of reliable evidence showing the treatments worked. Since Singh clearly believed the treatments were bogus and since the BCA was obviously promoting them, you might have thought the argument would hinge on how happy the association

---

[12] For those who want to know the specific science, the best solution turned out to be the so-called anti-k-T algorithm by Matteo Cacciari, Gavin Salam and Gregory Soyez (http://arxiv.org/abs/0802.1189).

was. Unfortunately, British libel law was a complete ass and it looked as though Singh, in order to defend an honestly made comment, was going to have to prove that the BCA was knowingly lying; that meant that so long as they could show that they didn't believe the treatments were bogus they would be home and dry. Finally, Singh persuaded the court of appeal that he didn't have to prove the BCA knew the treatments to be bogus, just that *he* had good reasons for contending that they were. The BCA withdrew its legal action at that point.

I went along to the meeting partly out of a sense of outrage that the law could be used to effectively silence a science-based critique, and partly just to have a beer with some friends who were going. Holborn is very near the UCL campus in Bloomsbury. I got a lot more out of this than beer. I met a whole bunch of intelligent, diligent and well-informed journalists working in newspapers and broadcasting, and I have to say this came as a bit of a shock. I met several excellent writers who wrote blogs and other online stuff. And I became aware of a set of people who knew and cared more about science and rationality than I, in my arrogance and ignorance, had expected to find outside the world of scientific academia I knew. This was as heartening, and as exciting, as the libel situation was depressing and dangerous. This wasn't just about Singh. Dr Peter Wilmshurst was being sued for analysing the health outcomes of a heart implant, and others have written better than I can about the chilling effect bad libel laws can have on scientific discussion when exploited by rich people or big companies. When Singh announced he was going to fight on, he appeared brave but unlikely to succeed. However, there was to be a campaign not just to help him, but to get the law changed, and there was at least some hope that it would be successful.

Some of the writers became good friends, and the campaign led to a measure of success in reforming the law – the new libel laws, which included some specific protections for peer-reviewed science, came into effect at the start of 2014. And Singh saw off the BCA. All of that was good, but looking back on this period, there was an additional and important benefit for me.

The campaign helped to break down some of the barriers that exist between the world of science, the world of academia (which overlaps with but is very different from the world of science) and other worlds, including those of the media, comedy and politics. This was a new experience for me, and proved rather important in influencing how I talked to people about what we were doing at CERN. Brian Cox, who was a friend and collaborator before he stormed those barricades with huge success, made it easier for us all by showing that if you present things well, people will be interested, but there was more to it than that. While I only participated in the libel-reform campaign in a very small way by writing letters and attending meetings, select committees and the like, I still made connections and gained an understanding of those other worlds in the process.

This removed a lot of my fear and suspicion of the media and politics, and also showed how lobbying could be done effectively. Along with a 'Science Blogging Talkfest' organised in 2010 by Alice Bell (then at Imperial, the large technical university in South West London) and Beck Smith of the Biochemistry Society, the libel-reform process did a huge amount to make me more comfortable with discussing science in public and via the media.

This permeability between different worlds seems essential to me, and while it requires some face-to-face interaction (and for me, at least, alcohol helps with this), online tools such as blogs and social media also have a very positive impact. It was after Alice and Beck's talkfest that Alok Jha, who was one of the panellists, asked me to join a new *Guardian* science-blogging initiative.

Exchanging ideas on Twitter in particular is something that has helped me a lot. Having a blog and a Twitter account that are read by a fairly large number of my colleagues in science and by contacts in the media turned out to be invaluable for my confidence when talking to bigger audiences elsewhere. Months later, I was sitting in a taxi on the way to be interviewed by John Humphrys on BBC Radio 4's *Today* programme about why the Higgs search was a waste of time. Knowing that if I said something stupid, or was misrepresented, I could at least tweet or blog about what I had really meant to say was a great comfort. As a scientist, it is a daunting thing to pop up briefly in the media and be given a few minutes, more or less at the mercy of professionals, to describe your work and why it is worthwhile. To have the means to reach people directly makes a big difference.

It turned out for me that in the particular example of the *Today* programme, John Humphrys was great and I didn't need to apologise for anything. But I still love that the tool invented by Tim Berners-Lee in the office downstairs from mine at CERN is supporting, amongst many other things, improved engagement between the science of CERN and the rest of the world.

Glossary: Bosons and Fermions

The word boson causes no end of trouble when people report on the search for the Higgs boson. It regularly gets spelled or pronounced 'bosun', and once, while being interviewed on TV and reading the autocue over the presenter's shoulder, I saw it spelled as 'bosom'. Like a true professional, Krishnan Guru-Murthy, the urbane Channel 4 News presenter, said 'boson' without blinking.

Boson is the name for a generic class of particles. The Higgs boson is one, but so are many other particles. In the Standard Model, all the particles that carry forces – gluons, the W and the Z, photons, plus the graviton, if there is one – are bosons.

Quarks, electrons and neutrinos, on the other hand, are fermions.

The difference between them is just spin. But in this context, spin is a quantum of angular momentum. It is a bit like the particle is spinning, but that is really just an analogy, since point-like fundamental particles could not spin, and anyway fermions have a spin such that in a classical analogy they would have to go round twice to get back to where they started. Quantum mechanics is full of semi-misleading analogies like this.

Regardless, spin is important. Bosons have, by definition, integer spin. The Higgs has zero, the gluon, photon, W and Z all have one, and the postulated graviton has two units of spin. Quarks, electrons and neutrinos are fermions and all have a half unit of spin. This causes a huge difference in their behaviour.

The best way we have of understanding fundamental particles is quantum field theory. In quantum field theory, a 'state' is a configuration describing all the particles in a system (say, a hydrogen atom). The maths is such that if you swap the places of two identical fermions, with identical energies (say, two electrons), then you introduce a negative sign in the state. If you swap two bosons, there is no negative sign.

Since swapping two identical particles of the same energy makes no physical difference to the overall state, you have to add up the two different cases (swapped and unswapped) when calculating the actual probability of a physical state occurring. Adding the plus and the minus in the fermion case gives zero, but in the boson case they really do add up. This means any state containing two identical fermions of the same energy has zero probability of occurring. Whereas a state with two identical bosons of the same energy has an enhanced probability.

This fairly simple bit of maths is responsible for the periodic table and the behaviour of all the elements. Chemical elements consist of an atomic nucleus surrounded by electrons. Because electrons are fermions, not all the electrons can be sucked into the lowest energy level around the nucleus. If they were, the probability of that 'state' happening would be zero, by the argument above. So as more electrons are added around a nucleus, they have to sit in higher and higher energy levels – less and less tightly bound to the nucleus. The behaviour of a chemical element – how it reacts with other elements and binds to form molecules, and where it sits in the periodic table – is driven by how tightly bound its outermost electrons are.

When bosons clump together they do some fascinating stuff too. The condensate that is responsible for the superconductivity in the LHC magnets, for instance. But it's hard to beat being responsible for the whole of chemistry, and therefore biology. And the rest.

Some theories extend the Standard Model by relating force-carrying bosons to matter-particle fermions. They do this by introducing a new symmetry between them. This symmetry is so mathematically compelling that it is called 'super' – supersymmetry – but that's another story.

1.7 Boost One

Something else that we had time for, while the LHC was being fixed, was a lot more study with simulated data.

To be honest, by now a lot of us were pretty sick of this. I'd left the analysis of real data when I left the ZEUS experiment in 2005, and had been doing little but simulation studies since. There's a joy in a well-written program, and a lot you can learn from modelling things in computer code. But without actual experimental results it all begins to feel a bit self-referential.

We did now have better software tools, some of which had been improved using information gleaned from the few 'beam splash' events we had recorded before the LHC broke down, and also by studying cosmic-ray data taken with the full ATLAS detector. Cosmic rays are particles from space that bombard the Earth constantly.[13]

And, very specifically, Adam and I could do some more work on our fancy boosted Higgs analysis. I mentioned this in section 1.3; now may be a good time to expand on it.

The *Colliding Particles* films I mentioned, made by Mike Paterson and featuring Gavin, Adam, me and a cast of dozens, were popular with teachers. They show a lot about how particle physics is done. As intended, they seemed to be doing a decent job of hitting the UK's ever-changing National Curriculum on 'how science works'. But even though they are

---

[13] See 2.2 Minimum Bias and 4.2 Science Board for more about them.

partly based around our scientific paper about one way we might find the Higgs boson,[14] they do not contain much actual physics. There was a review of the films in *Physics World*, the magazine of the Institute of Physics, the UK professional body for physicists, which gently pointed this out.

Fair comment, which I felt ought to be addressed. So I wrote my first ever blog post in an attempt to explain the physics behind the paper. Here is a rewritten version of that, which concentrates on explaining the new ideas in the paper rather than giving a summary of why the Higgs is interesting, which will come later . . .

What we knew at the point the paper was written was that if the Higgs boson existed, and if its mass was what seemed to be the most likely value (around 120 GeV[15]), then lots of Higgs bosons would be produced at the LHC. The difficult trick would be to pick them out from all the other things that would be going on.

The Standard-Model Higgs boson is a wave, or, more correctly, an excitation, in a quantum field[16] that fills the universe. Interacting with this field is what gives the other particles mass. Particles get different amounts of mass depending on the way they 'couple', or stick, to the field. This meant that the Higgs boson itself would couple to the mass of every particle, and would therefore be very likely to decay into the heaviest particle it could. These decays would happen super-quickly, so all we would see would be the things into which it had decayed. We would have to work out from them whether or not there was, briefly, a Higgs boson produced by the LHC.

For a 120 GeV Higgs, the heaviest thing it could decay into would be a pair of bottom (b) quarks. Quarks come in pairs. Down and up quarks are all you really need to make protons and neutrons, but for some not

---

[14] Referred to in the films as the 'Eurostar paper', as a sort of code for an initially underground London/Paris collaboration.

[15] GeV stands for gigaelectronvolts, the particle physicist's preferred unit of both mass and energy. I will say more about what they mean in section 2.1.

[16] See Glossary: Fields, Quantum and Otherwise (pp. 57–60).

fully understood reason we also have strange and charm quarks (similar to the down and up respectively, but more massive) and bottom and top. The strange quark was so named because the first evidence for it was particles with strange properties (basically living longer than expected and decaying oddly) seen in cosmic-ray interactions. 'Charm' seems a very whimsical name, but I guess the quark was quite charming when it turned up because it solved some tricky issues with the weak interaction. When the last two quarks were discovered, the names beauty and truth were proposed, I guess extrapolating from 'charm'. This shows you how unreliable extrapolation can be . . . Anyway, everyone calls them top and bottom now. Top only showed up at Fermilab, the US national particle physics laboratory in Chicago, and home to the Tevatron, in 1995. This was a great relief as it finally laid to rest the theoretical models in which the bottom quark had no partner, usually giving rise to tediously suggestive names – bare-bottom models, topless models, you get the idea . . .

I digress. The Higgs, if it were to have a mass of around 120–130 GeV, would not have enough energy to decay to top quarks, or to W and Z bosons, and so would mostly decay to bottom.

Bottom quarks also then decay (after travelling a few 100 microns or so) and each one gives a spray, or jet, of hadrons. We could see some of these particles in our detectors, and would be able to reconstruct the fact that two bottom quarks decayed, and that they came from a Higgs decay.

So: if a Higgs boson were to be produced, it would decay to two b quarks, and they would give two jets of hadrons.

The problem is, lots and lots of b quarks and jets of hadrons are produced at the LHC, and most of them have nothing to do with a Higgs. Before our paper, it looked like this background noise would completely swamp the signal and we would have to rely on other, rarer, Higgs decays to find it. Not only did this make finding the Higgs harder, but even if you found the Higgs some other way, seeing the decays to b quarks is actually pretty important in proving that whatever you might have found really was a Standard Model Higgs boson.

The idea we had in the paper was to look at those collision events where a Higgs is not just made, but is made and given a lot of kinetic energy – i.e. it is moving very fast, at speeds that are a substantial fraction of the speed of light. This happens in about 5 per cent of the Higgs production events we were simulating (assuming the LHC design beam energy of 14 TeV), so we would miss a lot of Higgs bosons by only looking at the fast ones. In fact, with the LHC starting at lower energies we would waste even more of them. But the advantage of this approach would be that we would lose still more of the backgrounds, the non-Higgs events, because the background jets usually have a lower energy.

Something happens when you look at these fast-moving Higgs decays. The faster the Higgs moves, the smaller the opening angle between the two b quarks it produces when it decays. In fact, very often the jets from the two quarks merge into a single jet.

This is a problem if, as had been done in all the studies until then, you look for two b-quark jets as your telltale signature that a Higgs boson was there. In our paper we turned this problem into an advantage. By looking at the internal structure of this single jet, at its substructure, we could see evidence for the two b quarks and the Higgs decay, get rid of even more background, and measure the mass of the Higgs boson well enough to make it stand out over the remaining background. What had looked like a hopeless case was recovered as a promising way of finding the Higgs at the LHC.

The idea of looking at jet substructure to find the decays of fast-moving particles had been put forward earlier by Mike Seymour,[17] and again in the paper[18] I mentioned in 1.2, which I wrote with Brian and Jeff. But Gavin had a better way of doing it. And together with Adam and Mathieu we were the first to apply these ideas to the Higgs boson.

The unexpected delay of a year meant that – with Erkcan Ozcan, a UCL postdoc, and a group from Freiburg – Adam and I could try it out on a

---

[17] See http://inspirehep.net/record/359650?ln=en.

[18] See http://arxiv.org/abs/hep-ph/0201098.

fully simulated ATLAS detector. We found it should still work pretty well even with a more realistic estimate of the experimental errors and a more complete study of the backgrounds.

A sign that these ideas were picking up fans was that out of the blue, Gavin and I received an invitation to go to the SLAC National Accelerator Laboratory to talk about them at a meeting called 'Giving New Physics a Boost'. Or just 'Boost'. This was July 2009, and there were by now quite a few papers coming out developing new ways of finding the decays of boosted particles and highlighting applications at the LHC. If you have enough energy in your collider, boosted heavy particles turn up quite often. So, even though London to SLAC is a long way to travel for a two-day meeting, it was worth the trip.

SLAC is in Menlo Park, California, about 40km from San Francisco. I'm not good at long-haul flights and the two days passed in a bit of a daze, to be honest, though from the discussions it was clear that exciting physics was happening, to the extent that a longer, follow-up meeting was already planned for 2010 in Oxford. I remember sitting on the seafront in San Francisco eating clam chowder from a bread bowl, waiting for the flight home and wondering what had just happened. Whatever it was, it was good.

Anyway, by the autumn of 2009 the LHC was repaired, at least to the extent that CERN was confident it could get the beams up to half the design energy – still a factor of 3.5 higher than any previous experiment. Slightly chastened but wiser, we were ready to try again for real data.

TWO

Restart

December 2009–March 2010

2.1 Low-Energy Collisions and Electronvolts

The LHC officially became a collider on 23 November 2009, when it began smashing together beams of protons with 450 gigaelectronvolts (GeV) of energy.

This is not a particularly high energy. For comparison, the Tevatron in Chicago was running at an energy of 1000 GeV. But it was the first time that we could measure real collisions in our detectors. The ATLAS detector recorded particles produced by a collision for the first time at 2:22 in the afternoon. All the experiments recorded data that day, and from that moment on, the LHC was a real physics experiment.

Those units of energy (GeV) probably aren't very familiar if you aren't a physicist. An eV is an electronvolt, and a volt is a measure of electric potential. So a standard domestic battery might provide a potential of 12 volts. If you allow a single electron to fall through this potential (that is, be repelled from the negative terminal and attracted to the positive terminal of the battery, because electrons have negative charge and like charges repel, opposites attract), it will pick up some speed.

Things that are travelling at speed have energy because of that speed, which we call 'kinetic energy'. The kinetic energy this electron will have gained due to the speed it acquires accelerating through the voltage provided by the battery will be 12 eV (12 electronvolts). That is how an eV is defined – the kinetic energy picked up by an electron as it falls through an electrical potential of 1 volt.

Since an electron is a very tiny thing with a tiny electrical charge, 1 eV isn't very much energy at all. The standard unit of energy is a joule, which is 1 kilogram (metre per second) squared (kg m^2 s^{-2}). The approximate expression for kinetic energy in terms of mass and speed is ½ mv^2, so if you have a kilogram mass (m) travelling at a speed (v) of 1 metre per second, it will have a kinetic energy of ½ x 1 x 1^2 = ½ a joule. It takes a lot of electrons to make a kilogram, so 1 joule is 6.24 x 10^{18} eV, that is, 6.24 million million million electronvolts.[19]

Clearly the eV isn't a very useful unit for everyday life (though you would lose weight very quickly, possibly terminally, on the electronvolt diet). It is handy for physics and chemistry, though, where we are often discussing the amount of energy needed to move individual electrons around. Chemical bonds often involve transfers of electrons with energies of a few eV, though it can be much more. To break up a water molecule, for example, you need to give an electron about 500 eV of energy. Also, when electrons move between energy levels in an atom or a molecule, they absorb or emit photons – quanta of light. In a sodium lamp, for example, electrons jump between levels separated by about 2 eV in energy, and give off the distinctive yellow light of sodium street lamps.[20] So a yellow photon has an energy of about 2 eV.

X-ray photons have energies of a few thousand eV (kilo-electronvolts, keV). These can knock even the most tightly bound electrons away from the atomic nucleus. To break up the nucleus itself, you need even more energy – millions of eV (mega-electronvolts, MeV). This is the energy domain of nuclear physics.

I said that unless you were a physicist, you might not be familiar with this. To be honest, unless you are an astro- or particle physicist, you

---

[19] If food calories mean anything to you, 1 calorie is about 4200 joules or about 2700 million million million eV.

[20] This behaviour is at the heart of spectroscopy, which I get to again in section 7.3.

probably aren't used to GeV (giga-, or billion, electronvolts) or TeV (tera-, or thousand billion, electronvolts). Single particles with this kind of energy only really occur either in high-energy cosmic rays or in big accelerators such as the Tevatron (you can probably work out where the name comes from now) or the LHC. With particles at these energies, you can not only break up the nucleus, you can shatter the protons and neutrons inside it. And, if nature works that way, maybe even shatter the quarks and gluons inside them.

So, a TeV is a lot of energy for one particle to have. But these particles are very small. If you took all the kinetic energy in a 1 TeV proton (say from the Tevatron) and gave it to a 1kg mass, the kilogram would barely move.[21] However, a proton with the same kinetic energy would travel at nearly the speed of light.

Enough about units. Something fun happened on Tuesday, 8 December 2009.

The disaster of 2008 had occurred when the magnet currents were ramped up high enough to bend 7 TeV proton beams. After the repairs, caution ruled and the plan was to take them up gradually to a level that could cope with 3.5 TeV beams. At 450 GeV (0.45 TeV) we had a long way to go. We would be watching carefully, and quite nervously, as the accelerator teams took the beams up in steps.

In the run-up to 8 December, the LHC team were edging the beams up in energy and got above the Tevatron record of 1 TeV per beam. So the LHC was now providing the highest-energy beams in the world, but had not yet brought them into collision at those energies. That Tuesday, there was obviously something going on with the beams, and the ATLAS shift crew were watching carefully. They even had parts of the detector switched on, though no plan for collisions had been announced. By coincidence there were a number of UCL people around – one student (Catrin Bernius)

---

[21] From the example above, 1 TeV is 10^{12} eV, so 1 joule is 6.24×10^{6} TeV, so you'd be giving the kilogram about 0.17 millionths of a joule. Invert the kinetic energy formula to calculate the speed and you can work out that the speed of the kilogram would be about 0.5mm per second.

on shift in the control room, plus Adam Davison and Nikos Konstantinidis looking at the 'event display', the graphic program used to display the results of a collision.

At 21:40 the two beams crossed and there were a few collisions, almost by accident it seemed. But ATLAS was ready – the shift crew recorded them, sent them through to the event-display people, who sent them on to Fabiola Gianotti, the head of the experiment. The first proton–proton collision to go above the maximum energy of the Tevatron! Not that it was any use for physics – the detector magnets weren't even on – but you could see particles produced from the highest-energy particle collision ever made in a laboratory, and that was the moment the LHC became the highest-energy collider ever. It was on the web page the morning after. A bit silly, but fun.

2.2 Minimum Bias

In March 2010 the first physics papers based on LHC collisions came out. They addressed the first thing we needed to do, which was to measure the properties of the average, or 'minimum bias', collision.

It is surprisingly difficult to define a collision.

Two protons approaching each other in the LHC will repel each other because they both carry positive electric charge and things that have the same charge repel each other. This electromagnetic force falls off with the inverse square of the distance,[22] so if the distance between the protons doubles, the force decreases by a factor of four. But it never goes to zero. So even two protons that miss each other by miles (OK, microns inside the LHC beam pipe) will repel each other slightly and bounce away from each other a tiny amount, and so could be said to have 'collided'.

In practice, we would not detect such glancing collisions. The protons

---

[22] That is, $1/r^2$ if the distance between them is r.

would carry on down the LHC beam pipe and would never make it into the ATLAS detector.

Protons that come closer to each other, however, can undergo much stronger scattering, and of course in many cases they smash each other to pieces. At such short distances and at such high energies, the electromagnetic force is, in fact, not the main effect; most interactions between the protons are carried by the strong nuclear force – QCD, the theory of quarks and gluons.

The strong force really is short-range. It barely reaches beyond the size of an atomic nucleus. But where it reaches, it is *strong*. It is strong enough to overcome the huge electromagnetic repulsion between the two positively charged up quarks inside a proton, for example.

Many of the strong collisions are still glancing blows, though. And in most of them the protons hardly get broken up, or don't break up at all, and mostly they don't register in ATLAS. These are called diffractive collisions. There are some specialised detectors, tens or hundreds of metres away from ATLAS or CMS down the LHC beam pipe, that can pick up some small fraction of the unbroken protons from these events in special LHC runs, but most of them get overwhelmed in normal LHC running.

So when do you say there has been a real collision? How do you define that? It matters, because we try to measure things like the average particle distributions in so-called 'minimum bias' events. What are we averaging over? On the face of it we would try for an unbiased selection of collisions. But we can't *possibly* be truly unbiased; since the vast majority of glancing collisions don't leave any trace in the detector, we can't select them, whatever we do. In practice we will see most of the non-diffractive events – where the protons are smashed up and some of the bits hit the detector – and only see a few of the diffractive events.

Here's an analogy. Say you want to measure the average height of people in the UK.

You go out at lunchtime every day and pick a random selection of people off the street and measure their height. You add up all the heights and divide by the number of people you measured, and you have the average height. Just like us counting how many particles are produced in all the collisions we see, and dividing by the number of events seen. Simple. And wrong.

The problem in both cases is that the people (or collisions) we see constitute a biased sample. In the collisions, we miss the diffractive events, in which fewer particles are produced, so we will overestimate the average number of particles produced in a collision. In the height case . . . well, you are measuring at a time when most children are in school.[23] And children are shorter than most people. So you will overestimate the average height because you haven't accounted for them.

You could do one of a few things here to try and improve matters.

1) You could, for example, make an estimate of how many children there are in school, so how many you would have seen if you had had a chance of measuring their height; you could model (estimate, guess) what their distribution of heights might be, and you can put all that together to apply a correction to the data.

Or 2) you could say, I'm just going to measure the average height of adults. So in my random sampling, I will not measure the height of any of the children I meet who happen to be bunking off school.

Or 3) you could say, I'm going to measure the average height of people on the street at lunchtime. This way, by definition you've got the right answer already, so long as you had no other biases in there.

Historically, experiments have used theoretical models, either to add in the missing diffractive events (the kids in school – first option) or to remove the small remaining diffractive contamination and produce measurements of what they call non-diffractive events (adults only – second option).

The trouble is, this means that what they measured is only defined within a particular theory. 'Diffractive' and 'non-diffractive' are really just

---

[23] Unless you are measuring in the school holidays. Or on a weekend. I am a bit disturbed that I feel compelled to add this footnote.

words. There's no clear, unambiguous definition of the separation between the two. In the UK, we choose the 18th birthday as the somewhat arbitrary but universally accepted dividing line between children and adults. There is no such widely accepted division for the separation between diffractive and non-diffractive in physics.

If you use a model to correct for diffractive events, you buy into a particular theoretical definition of them, and a set of assumptions as to what they look like. You are no longer just reporting what happens. I think it is very important that, having gone to the enormous trouble of building the LHC and the detectors, we first just measure what happens, with the minimum of theoretical input. The next step is of course to confront the data with theory as part of the process of exploration and understanding. But the first, reporting, step is essential.

Fortunately my collaborators agreed. So instead of measuring the particle distributions in 'non-diffractive' events, or averaging over all events, ATLAS measured the distributions for all events that have at least one charged particle in a given region, regardless of whether a particular model would call the event 'diffractive' or not. This is like option three above: measuring everyone you meet, adult or not, but stating the sample bias explicitly. The charged-particle requirement is a physical criterion that can be reproduced regardless of any model (it is the equivalent of 'meet them on the street'). The difference is significant (up to about 20 per cent). But the difference in *principle* is huge, and comparisons to models become much less ambiguous.

This approach was surprisingly new, and in some quarters controversial. Partly because of conservatism – 'That's not how we did minimum bias on my experiment' – and also perhaps because 'average height of people I met' is not such a useful number as 'average height of people in the UK'. But it's the only thing we can actually measure – the rest is interpretation.

These measurements, and more like them since then, have told us various things. One of the main things was that the detectors were working well. But more than that, they tell us about QCD and the proton. We know the proton is full of quarks, stuck together by gluons, the force carriers of QCD. But while we can write down what we think is the fundamental equation governing this, we have not solved that equation for the proton, and neither had we measured its behaviour at these energies before. Because the fundamental equation is not solved, there are lots of models, based on various approximations, with various free parameters in them. The new data allowed some of these models to be rejected, and allowed the parameters of some of the others to be constrained and improved.

Apart from the intrinsic interest of understanding these things, we need this better understanding of 'average' events in order to more effectively find rarer events amongst them. Events in which photons, W or Z bosons or top quarks were produced, for instance. Or even Higgs bosons.

Another place these improved models are needed is in understanding the showers of particles produced when a proton, or some other particle, hits the upper atmosphere. Super-high-energy particles, with energies much higher than the LHC, bombard the Earth all the time.[24] They collide with atoms in the upper atmosphere, smash them to pieces, and the result is a spray of particles that can be detected at the Earth's surface by experiments such as the Pierre Auger Observatory in Argentina. The more we understand about the processes involved in these high-energy collisions, the more confident we can be in using such detections to work out the energy, direction and composition of the initiating particles, and hence where they come from and what violent stuff is going on out there in space to produce them.[25]

None of these first papers were the really exciting new physics the LHC had been built for, but it was interesting and important for firming up our knowledge as we stepped into the unknown. And it was great to have real data at last!

---

[24] This is perhaps the most compelling evidence that such collisions don't cause black-hole/death/horror catastrophes, since we are all still here at time of writing.
[25] I like to think they are crossfire from interstellar space battles at the fall of the First Galactic Empire.

2.3 Energy and Mass

During this time I was also teaching a course to first-year undergraduate physicists at UCL. It was a lot of fun (for me, anyway) being able to drop the latest news from the LHC into the start of a pretty dry lecture on differential equations or matrices.

The course is fairly typical of first-year physics degree courses; the main goal is to provide students with the mathematical tools they need to do degree-level physics. It contains some techniques for solving differential equations and for doing multidimensional integrals, matrix manipulation and coordinate transformations. The sweetener at the end is that we do Einstein's special relativity.

When I was talking about the energy of the LHC beams at the start of this chapter, I didn't give the speeds, apart from to say 'nearly the speed of light'. That's because I can never remember how many nines to put after the decimal point. This is clearly a great failing of mine. I was asked the speed of the protons in the LHC by an audience member after a talk at the headquarters of the huge Wellcome Trust medical charity once, and they seemed very miffed that I didn't know the exact answer.[26]

The answer is 'nearly the speed of light', of course, and that has been the case for every high-energy accelerator for decades. The more precise answer was 0.999999964 times the speed of light in 2010, and on reaching full energy this will be 0.999999991 times the speed of light. In the first case this is 299792447 metres per second and in the second 299792455 metres per second. So all the effort in 2013–2014 to get up to full energy buys us another 8 metres per second in speed; about as fast as I cycle to work.

This illustrates the fact that it is actually the energy that matters. Special relativity means that protons can never reach the speed of light, even though they can keep on getting more energy and momentum. In everyday physics, where speeds are small fractions of the speed of light,

[26] I did get gingerbread Daleks (for fans of Doctor Who), though, which was a bonus.

momentum is mass multiplied by velocity. In special relativity there is a gamma factor (that's γ) that goes in front of this, which is nearly one at low speeds, but gets very large as the speed gets close to the speed of light.[27] The gamma factor is a consequence of Einstein's fundamental postulate in special relativity, that the speed of light is the same for all observers. So the momentum gets very big even though the speed only gradually approaches the speed of light. The same thing happens with energy; the total energy of a particle is actually $E = \gamma mc^2$. When the speed is zero, $\gamma = 1$ and this is just the famous $E = mc^2$. When the speed, v, is much less than the speed of light, c, this is approximately $E = mc^2 + \frac{1}{2} mv^2$, where the second term is the kinetic energy. But at speeds near c, γ gets huge and that approximation doesn't work any more. The energy can keep on increasing but v can never quite reach c. It's all a bit odd, but it works.

Particle physicists very often talk about masses and momenta in units of GeV or MeV. Strictly speaking this is wrong. Energy is in GeV, momentum is GeV/c and mass is GeV/c^2. We avoid having to scatter factors of c all around the place by using what we call 'natural' units, where c is defined to be 1. In these units, Einstein's equation is even easier: $E = m$.

The mass of a proton is about a GeV.[28] So a proton in the LHC travelling with an energy of 4000 GeV is very highly relativistic, by which I mean that its kinetic energy is more than 4000 times bigger than its rest mass energy. This implies that when we collide two of these protons head-on, the energy that is in principle available to make new particles is 8000 GeV. This is enough to make more than 8000 new protons out of the collision of just two. And of course this energy might make something much more interesting, and newer, than that . . .

---

[27] This isn't a textbook, but I can't resist giving the actual expression for γ, which is $\gamma^2 = 1/(1-v^2/c^2)$, where v is the speed of the particle and c is the speed of light. You can see from this that as v gets close to c, γ approaches 1/0 (one divided by zero), which is infinite.

[28] 0.938272046 +/- 0.000000021 GeV, actually.

2.4 'Is There Any Chance You are Going to Destroy the World?'

That's one of those tricky questions scientists hate being asked.

It is very hard to prove a zero probability, especially if you take quantum mechanics seriously, which I do, of course. So conversations can often go like this . . .

'Is there any chance you are going to destroy Geneva/the world/the universe?'

'No significant chance, no.'

'No significant chance? You mean there is a chance, then?'

'Well, possibly an insignificant one. But—'

'You evil bastard! Any chance at all is too much! I like their chocolate!/I have kids you know!/Wow that's kinda cool, but . . . !'

'Wait. I said—'

'You people must be stopped!'

No one leaves that discussion feeling good. The next interview is likely to go:

'Is there any chance you are going to destroy Geneva/the world/the universe?'

'No.'

'Are you really sure?'

'Yes. Shut up, you scaremongering idiot.'

Which doesn't really advance the public understanding of science, and may in fact be less reassuring than the first approach. Some way down that route lies: 'Of course there's no risk from mad cow disease. Look, I'm feeding a burger to my child . . .'

The problem is a misunderstanding of probabilities and, possibly, of significance.

Once upon a time I went to a festival (the Secret Garden Party) to give a talk about the LHC for a group called Guerilla Science that arranges such things. I took with me my son, who was then aged seven. We saw some bands, bought some hats and camped overnight. Then we got up and listened to some talks in the 'Science and Rationality' tent, which was where I was going to give my own show. One of the speakers was David Spiegelhalter, Winton Professor of the Public Understanding of Risk in the Statistical Laboratory, University of Cambridge, and very good at explaining risk and probability.

Spiegelhalter described the idea of a micromort: a term invented by Ronald Howard for a one-in-a-million chance of death due to a given choice or action. Eating a burger, smoking a joint, crossing a road . . . any given action can, given enough data, be assigned a risk of death in terms of micromorts. Getting up in the morning carries micromorts. So does staying in bed. Part of the context was the so-called 'Equasy' affair, in which the British Government's drug adviser Professor David Nutt had pointed out that taking the drug ecstasy carried about the same risk as horse riding (about 0.5 micromorts a hit).

Micromorts and balance of risk made a great impression on my son. More than a year later, he had to discuss for his homework the question 'Why do people take risks?' His answer was: 'They have no choice. All you can do is decide which risk to take.'

Now my son is an utter genius, of course, as is David Spiegelhalter. But the rest of us can still learn to understand risk better than we do. Perhaps some scenarios will help.

Imagine doing something for the first time. Say a new experiment. Say, oh I don't know, the LHC, or RHIC,[29] or the Tevatron, or one of the previous machines some people have got agitated about. Then take the worst consequence imaginable, even if it is in contradiction to all experimental evidence, theory and even logic. As science has a hard time proving a negative, you might conclude that there is an infinitesimally small chance of the bad thing happening and be inclined not to go ahead. But before deciding, you have a duty to consider also the risk of not doing it.

---

[29] Relativistic Heavy Ion Collider, Brookhaven, NY, USA.

Scenario 1

The year is AD 2125 and the Earth has a serious problem. A rogue planet, drifting through the spiral arm of our galaxy, has just been detected via several innovative astronomical techniques involving gravitational wave detectors and observations from our deep-space telescope system. The planet is on course to enter the solar system and is massive enough to seriously disrupt the orbits of the other planets. Many-body quantum gravity calculations, using the detailed knowledge we have accumulated of the masses and trajectories of our 'local' environment, indicate a near certainty that one result of this disruption will be to send the Earth crashing into the Sun within two decades. Fortunately these observations and calculations have given the inhabitants of Earth enough warning. Using the latest antimatter fuel cells, a small robot ship is sent to the planet. Once on the planet, nanobots assemble a mini-black-hole factory that is used to provide a small but steady and powerful warp drive, diverting the planet away from the solar system into a cosmological near miss. Party time.

Scenario 2

The year is AD 2145 and the Earth has a serious problem. There's no one to help it because its most intelligent species wiped itself out in a nuclear armageddon/global climate catastrophe/whatever the next one is. The Earth mostly bounced back from this and is still a marvellous cradle of life, until a rogue planet sends it spiralling into the Sun in the mother of all environmental catastrophes.

Scenario 3

The year is AD 2135 and the Earth has a serious problem. A rogue planet, drifting through the spiral arm of our galaxy, has just been detected. The planet is on course to enter the solar system and is massive enough to seriously disrupt the orbits of the planets. Calculations using all the knowledge we have of the masses and trajectories of our 'local' environment indicate a significant probability that one result of this disruption will be to send the Earth crashing into the Sun within a few years. Unfortunately, many commentators and politicians refuse to believe this and claim the whole thing is a left-wing (or right-wing, according to taste) conspiracy. Anyway, there's not a lot we can do about it. The warning has come a bit late, and we still do not even know whether there is a Higgs or extra dimensions or mini black holes, so we have no suitable power sources to get us to the planet and no way of dealing with the threat even if we did. We spend five years moping about the foolishness of the 'safety first' legal ruling that made us turn off the LHC back in 2010, as well as the similar rulings that followed on new experimentation across many fields of physical and life sciences. Then we crash into the Sun.

The above scenarios are of course three amongst an infinite number of tiny possibilities. However, anyone who advocates stopping research because of imagined doomsday scenarios should also be made to estimate the risks associated with stopping the research and the doomsday scenarios we might thereby be exposed to.

2.5 Impact

Implicit in the scenarios just described is an assumption that knowledge practically always turns out to be useful, often in unpredictable ways. Or, if you prefer, knowledge is power.[30]

For the whole of my career as a scientist, there have been arguments about whether research, and research funding, should be directed towards known benefits (economic, medical or others) or somehow 'purely' for the joy of finding stuff out. These arguments sometimes degenerate into

---

[30] Often attributed to Sir Francis Bacon. It sounds even better for my purposes in Latin – '*scientia potential est*' – though what Bacon actually wrote was '*ipsa scientia potestas etc*' (1597), meaning 'knowledge itself is power'. Which is also a good thought, but now the Bacon and potato thing is making me want tartiflette, which serves me right for spending too much time on Wikipedia.

partisan battles between various interest groups, but usually they are between two sides, both of whom believe strongly that scientific research is a societal good, even a necessity. The disagreement is over the best strategy to optimise the benefits.

Most participants in the debate agree, for example, that encouraging more young people to study physics is a good thing, because physicists are economically beneficial things to have around.

You might think – I do – that the excitement of the LHC and the wonder of reaching further into the heart of matter than ever before would be a major positive in that regard. Physics stories don't often get into the news on the scale that we did on 10 September 2008, for example. But Sir David King, former Government Chief Scientist and at the time President of the British Association for the Advancement of Science (BAAS, now the British Science Association), felt so strongly that research should be directed at urgent practical problems and applications that he went on *Newsnight*, on the evening of the biggest physics story in decades, to accuse us particle physicists of 'navel-gazing'.

Luckily for physics, Brian Cox was on hand. He wasn't then as well known as he is now, and was quite nervous and wound-up after what had been an amazing day of high-stress physics in public.[31] Even so, he was thankfully more confident with the media than any other member of an LHC collaboration, so we did not have to end the most exciting day of physics for years with a smug, unchallenged takedown on prime-time television. After a moment's unfeigned total astonishment at the 'navel-gazing' accusation, Brian managed to make some effective arguments about the benefits of particle-physics research in terms of both its direct spin-off technologies and its inspirational effects, which were obvious on that day of all days. The objectives of the BAAS – 'promoting science, directing general attention to scientific matters, and facilitating interaction

---

[31] Initially, when Brian was on the way back from CERN for the programme, *Newsnight*'s planned opponent for the ritual confrontation had been the right-leaning lobby group the TaxPayers' Alliance. Sir David was probably more interesting.

between scientific workers' – were in the end served, though I don't think much credit went to its president.

Anyway (and you can probably tell the sense of betrayal still rankles . . .), although that moment was spectacularly poorly chosen, the argument has not gone away, and nor should it. The conflict between applied and 'other' science is so artificial and damaging that I have trouble deciding what the 'other' should be called. Not 'fundamental', definitely not 'pure' (there's no such animal). 'Curiosity-led' is about the best I can reach for. I think that when we study far-off galaxies or the Higgs boson we are mostly motivated by curiosity as to how the universe works. When we study new materials, or climate, we may be primarily motivated by trying to solve a pressing problem or develop a cool new technology. But the boundaries are fuzzy: the vast majority of scientists working on the LHC would be overjoyed if applications were found for our discoveries, and many of the software, detector or accelerator developments lead physicists off at highly applicable tangents. Likewise, while the desire to cure disease or save the world from climate change is a serious and rational priority for many, most scientists I know in such areas have a deep intellectual curiosity about the fundamental mechanisms they are studying.

At the moment I am head of the Department of Physics and Astronomy at UCL. Actually, while I am working on this chapter right now, we are going through an exercise known as the Research Excellence Framework (REF[32]). I am the 'unit of assessment lead' for three departments: mine, the Mullard Space Science Laboratory, and the London Centre for Nanotechnology. The range of science and engineering done in these departments covers cosmology, quantum computing, particle physics, materials science, the physics of biological systems and much more. We have been required to collect evidence of how our work has affected society beyond academia, and the case studies come from all these areas.

---

[32] Personally I expect this clunky name is another acronym-driven contrivance, as in the cries of 'Oi, Ref!' heard both in universities and football grounds all over the UK.

One of those making the big jump is precision-lens manufacturing (for astrophysical projects such as the Dark Energy Survey, very 'curiosity-led' cosmology), which is now at the heart of a company started by Professor David Walker in North Wales, selling machines and expertise to a wide range of manufacturing and engineering companies. The technology they have developed is used to grind and polish everything from precision components for the space industry to replacement hip and knee joints for health care.

The purpose of the above paragraph is not to bang the drum for UCL. Any major physics department could paint a similar picture (and in the UK they all are, at the moment, thanks to the REF). The point is that for people like Sir David King to pick out a part of this ecosystem as useless (which is what I take 'navel-gazing' to mean) is dangerous nonsense. Even though the lead times from basic research to application can be very long, there are already examples of technology developed for the LHC that are being deployed in practical situations. Novel detectors developed for the LHC are being used to monitor radiation doses for hospital patients undergoing radiotherapy, and are also being developed for retinal implants. The worldwide computing grid technology, put in place primarily to analyse LHC data, has been used to run structural analysis for antimalarial drugs.[33] Physics is useful and interesting all over the place, and this is why many very bright young people want to study it, many of whom take their degrees and go and do amazing things outside of academia or physics. But I do not ever want to be in the position as a senior physicist of having to tell the new students at induction that they should 'learn all this stuff because it's useful, but don't get carried away trying to do new fundamental physics, we don't do that here'. Any more than I would be happy telling them that string theory is the top of the pile and they should avoid getting their hands dirty with data and applications.

---

[33] See 'Particle Physics, It Matters', Institute of Physics, 2009, http://www.iop.org/publications/iop/2009/page_38211.html.

2.6 From Liquid Argon to M-Theory

Finally, before moving on to the first high-energy running of the LHC, I want to say something about the relationship between experiment and the multidimensional fringes of theoretical particle physics. Those fringes include string theory, in which fundamental particles are tiny vibrating strings, brane theories, in which they are higher-dimensional objects, still vibrating, and M-theory, which is a rather vague attempt to unify these approaches. In general, such highly mathematical frameworks strive to bring together quantum field theory and general relativity and grope towards a theory of everything.

Towards the end of a series of interviews with would-be undergraduate physicists who all want to be string theorists, I sometimes find myself sympathising more with the David Kings of this world. I do occasionally breathe a sigh of impatience and resist the urge to say, 'But I thought you wanted to be a real physicist?'

Such a slur on my theory colleagues would, of course, be unworthy. In the ecosystem I described in the previous section, theoretical physics is a vital component, and string theory and its relations are a legitimate part of that. Our observation and understanding of nature have led us to a number of beautiful and highly mathematical underlying principles. Using what are effectively thought experiments to see where these principles lead us, where they break down and what might replace them, is thrilling science. But for me, at least, it has to retain a connection with data.

I care much more about the ability of a theory to stand up to experimental testing than I do about its mathematical beauty. M-theory, string theory and so on have a long way to go in this regard. Even supersymmetry, a necessary prerequisite for string theory, struggles, although subspecies of supersymmetry are at least in contact (indeed, often conflict) with data, and will be discussed later in this story.

The LHC experiments may discover supersymmetry, and (or) find evidence for extra dimensions. This is really exciting, and either discovery would certainly add credibility to M-theory, as well as revolutionising our understanding of fundamental physics. But finding such evidence is the main thing, and the search tends to keep your feet on the ground even if the goals are highly abstract. Not only do we need the accelerator, we need our detectors to record what happens when protons collide. We also need to know how to interpret what our detectors are telling us.

Most high-energy particle-physics detectors surround the point at which the particle beams collide, and consist of concentric layers of different technologies, each designed to tell us something different about the collision. In one of the world's least helpful analogies, I have often described this as like 'a sort of high-tech cylindrical onion'. Various layers of this onion will crop up during the story, but one of the most important is the calorimeter.

This measures energy – calories. As discussed,[34] food carries energy, and particles carry energy when they are produced in a proton–proton collision at the LHC. We want to know, as precisely as possible, how much energy they carry, as this is vital for working out what actually happens in the collisions.

The basic idea of a calorimeter, whatever technology it uses, is to stop the particles with some very dense material. They hit the material, slow down, and as they do so they give off electromagnetic radiation (photons – light, basically). The amount of light given off has a correspondence with the energy the particle had in the first place. So the trick is to measure the light, work out the correspondence, and so measure the energy. Working out the correspondence is called calibration. It's difficult.

The main calorimeter technology in the ATLAS detector uses liquid argon interleaved with lead or copper to stop the particles. Liquid argon may sound exotic, but we use it because it gives out light in a very nice way, proportional to the energy going in. It is actually quite cheap, too: less per litre than a well-known brown fizzy drink. Plus it's very stable – the amount of light it gives out per GeV doesn't change with time – and it is resistant to radiation.

---

[34] See 2.1 Low-Energy Collisions and Electronvolts.

Even so, it does give off a different amount of light when it is hit by an electron compared with when it is hit by, for example, a pion (a commonly produced hadron made up of a quark and an antiquark). And the mix of pions and electrons hitting the calorimeter is not something we know in advance. There are many such subtleties you have to calibrate for before you really believe what you see.

The whole calibration process is equivalent to checking your experimental set-up in any small science experiment on a laboratory bench. It consists of many, many 'control' studies – measuring stuff you already know, in order to understand your apparatus. Without such procedures, we cannot believe any particle measurement we make, whether it's a couple of particles produced in a hospital scanner, or a spray of particles indicating extra dimensions at the LHC.

If and when the experiments start seeing well-understood signals over and over again, for supersymmetric particles or extra dimensions, that's when those theories win. And that's when, maybe, I'll start taking M-theory more seriously.

Glossary: Fields, Quantum and Otherwise

Commonly used words are often appropriated by areas of specialism and used to refer to something more specific and technical than their everyday meanings. An example in physics is 'work'. If a constant force acts on a particle and pushes it through some distance, the work done is defined as the force (in the direction of movement) multiplied by the distance travelled. It is a very specific quantity, and in fact is a form of energy. The energy of the particle will increase by the amount of work done. There is clearly a relationship between the everyday idea of work – making an effort to get something done, often for some financial reward – but the physics usage is precise and limited whereas the everyday usage is a bit fuzzy and very general.

Momentum seems to be a word that went the other way. In physics it is γmv (the relativistic gamma factor multiplied by the rest mass multiplied by the velocity), and is a way of quantifying the tendency of a particle to keep moving at a given speed in a given direction. And for speeds much lower than the speed of light, γ is very close to 1 so can be ignored. In more general usage momentum is used to describe the impetus behind a political campaign or some other social or political process or policy, with the same implied meaning that the more momentum something has, the harder it is to stop, but with no precise definition of what the momentum is in each case.

One of the words I have so far tried to use infrequently, but which I will really need to use more often in the following chapters, is 'field'. In general usage a field is a flatish bit of land with stuff growing on it, probably looked after by a farmer and possibly containing cows. It can also mean an area of study or expertise, and looking back I see I have already used it in this sense too. These meanings can also be mixed up – as in the reason the scarecrow got tenure.[35]

In physics, 'field' has a more technical, but related, definition. A field in physics is simply a quantity that has a value at all points over a certain region of space. So if you are in a room, there are various fields you might use to describe your environment. As a physicist you might do it this way:

First you need a way of specifying any given point in the room. A good way of doing this would be to choose the floor level at one corner of the room to define an 'origin'. Any point on the floor can be reached by going a distance (call it x) along the floor parallel to one of the walls that meets at the corner you chose, then another distance (call it y) along the floor parallel to the other. Then any point in the room can be reached by going up some distance (call it z). Three numbers, x, y and z, are all you need.

You can now talk about various useful fields. The temperature, for

---

[35] He was outstanding in his field.

example, has a value at every point in the room. Say on average it is the nominal 21°C room temperature. If it were really this temperature everywhere, then you would have a constant field, with no dependence on position in the room, so no dependence on x, y or z.

However, it is quite likely that the temperature is slightly higher near the ceiling than at the floor, since hot air is less dense than cold, so will rise to the ceiling. We could describe such a dependence with a field like $T(z)$, that is a value T depending only on the height, z. T would be a function[36] of z, maybe something like $T(z) = 20.5 + 0.5\,z$, with z given in metres and temperature in degrees Celsius, for example. For a 2m high room, the floor temperature would be 20.5 + 0.5 x 0 = 20.5°C, and the ceiling temperature would be 20.5 + 0.5 x 2 = 21.5°C. And the temperature at every point in between can be worked out from the expression for the field. Other fields could be used to describe the air density, for example, or even the amount of noise.

All of those are fields that are described by a single number at any point. They have a size, but they have no direction. We call them 'scalar fields'. A 'scalar' is just a word for a thing that has size but no direction.

There are other kinds of fields that have direction, too, and we call these 'vector fields'. I have already used examples of these – electric and magnetic fields such as those in the LHC magnets, for example. In the room there will be a gravitational field. It has a size at every point (a force of about 9.81 newtons per kilogram) and a direction (it points downwards).

Although the fact can be ignored for most everyday purposes, the electric and magnetic fields are in fact quantum fields.[37] This means that if you study them at very short distances, you will see that the field does not have a continuum of possible values, but is described by the

---

[36] 'Function' – there's another common word, appropriated by mathematics this time.

[37] Gravity may be too, but we don't know the theory yet.

sum (superposition) of a series of discrete[38] quanta, or excitations,[39] of the underlying quantum field. These excitations are a bit like waves and a bit like particles. The quantum field theory of electromagnetism, QED, has two fields – the photon field and the electron field. What we measure as electromagnetic waves, or as individual photons or electrons, are excitations in these fields.

Quantum or not, the idea of a field remains the same. It is a physical quantity that has some value, or superposition of values, at all the points in space over some region you are interested in.

---

[38] Discrete meaning discontinuous and distinct, rather than politely done in pastel shades.

[39] Another appropriation, with an obvious relation to everyday usage, since quantum field theory is very exciting.

THREE

High Energy

March–September 2010

3.1 Seven TeV

On the morning of Tuesday, 30 March 2010, the LHC really became the highest-energy particle collider in the world.

It began to accelerate protons to an energy of 3.5 TeV (3500 GeV), three and a half times higher than the previous record held by the Tevatron at Fermilab in Chicago, thereby breaking new ground in fundamental physics and accelerator technology. We were also breaking new ground in terms of doing science live and in public, or so it felt, at least.

We had hoped for collisions for breakfast, but it didn't work out that way. Although the false starts reminded me of frustrating night shifts on the ZEUS experiment in Hamburg where I did my PhD, they aren't unusual on a new collider. Given the major accident in 2008, however, plus the huge public interest, it was a very nerve-wracking morning. I guess this is how space scientists feel when launch delays happen in mid-countdown. Injecting and ramping the beam energy feels like the initiation of a countdown sequence and we had to go through it three times before we finally launched the high-energy physics program of the LHC. In time for lunch. Good enough. Fantastic, in fact.

I was in London, not CERN, and seeing how crowded the ATLAS control room looked, that seemed a good choice. Up to a hundred people really had something vital to do in the early running of the ATLAS detector at any one time, but there are 3000 people on the collaboration, contributing skills ranging from software and electronic and mechanical engineering to theoretical physics. Trying to cram everyone into the control room for the first collisions would have been impossible. A good thing, therefore, that CERN had invented the web, so there was a live feed and the results were available instantaneously. There was also a flood of emails, phone calls and tweets.

The storage and acceleration up to 3.5 TeV of the hair-thin proton beams in the LHC was achieved as planned. The beams were then brought into head-on collision and the results were recorded. We were at only half the design energy of the LHC, but this was still more than enough to break into the territory of unknown physics. The plan then was to do this routinely for a couple of years, collecting enough data to explore the new landscape thoroughly, with a particular focus on searching for signs of the Higgs boson. If the boson was there to be found, it might possess any one of a range of different masses, and might decay to a number of possible less-massive particles. Those decays would be relatively rare, and the signs that they had occurred would have to be sifted out from a range of different background processes that threatened to swamp them. So we needed to collect many collisions before anything was likely to be clear. Then we planned to pause, upgrade, and redouble the energy.

I seemed to spend much of the day on radio and TV news. After my nerves watching the LHC operators coax the beams into collision, and hoping ATLAS would successfully record the collisions, being interviewed on live TV was (almost) a breeze.

I remember hoping at the time that seeing a science story like the LHC develop in public over a few years would help people gain a clearer view of scientists, and also of the erratic but real process of science. The experience of the 2008 start and failure had given me some reason to hope. Scientists periodically pop up in the headlines of some story or other as either angels or demons and without much context,[40] then vanish

---

[40] Not even so much as a turgid thriller about blowing up the Vatican.

again. 'It's a breakthrough!' – but then it proves a bit more complicated than that and no one can be bothered to follow it up. The LHC story, however, seemed to be big enough that a subset of the media and the public were ready to stick with it through several twists. That could be good for particle physics, and maybe even helpful for other big science stories, including the more controversial ones, as well as those with a more immediate impact on our well-being.

3.2 This Is Not a Drill

In 2010, the UK general election was approaching. We'd had a terrible time with science funding around 2007–08 when, after a longish period of stability, the government used a merger of two research councils to make massive cuts in astronomy and particle physics. Then the global financial crisis began and science funding had been cut across the board.

In 2010 the financial crisis hadn't stopped and banking malpractice stories were still emerging regularly, but in the world of the research councils (the main bodies responsible for distributing taxpayer funds into research projects) things had settled into some kind of stand-off, with funding slowly declining but with enough resources around that we could still be major players in the LHC. The LHC was of course much more interesting than political lobbying, so I wasn't always paying as much attention to science policy in the election as I should have been.

This is fairly typical scientist behaviour, by the way. We just get the data and are happy. However, in many discussions with civil servants and those who work with them, I've heard that our big mistake was that in the economic good times, when science was getting some share of the growth (and particle physics was getting somewhat less, but not being cut), we had just shut up and got on with the science. The argument was that we should have been aggressively pushing for more and more cash, just so that there was more fat to cut when the bad times came. Sadly, this illustrates one of the gulfs in understanding between the worlds of science and politics, and it's an issue that impedes funding of science programs around the world, not just in the UK.

Research is about finding good questions to ask, then answering them. Good questions that can actually be addressed don't come along too often, and the capacity to identify and answer them grows slowly – you can't just go out and double the size of a good research group overnight. There are some big, wonderful and expensive things we could do if we had the money, but unless such an injection comes on the back of sustained, long-term investment, it would not take us long to run out of the skilled people to deliver them. In a world where business and other interests continually lobby, it's not always easy for politicians to place the right value on stability and the strategic planning of research capacity.

Scientists, on the other hand, mostly want to do science, and many of us only get noisy and politicised when our capacity to do science is under imminent threat. As well as being a huge tactical mistake, this is an insult to the interested and intelligent public who foot the bill and deserve to hear what we are up to. It is also a recipe for political and social disaster – public life needs science, and scientists who engage consistently. We should have been more engaged – at least saying thanks for the resources and reporting the good things they were funding.

But oh, the temptation of real data! Shortly after the high-energy start-up, Oxford University hosted the second 'Boost' meeting. This took the topic of jet substructure a lot further,[41] but the results were still all based on simulation, and the tension induced by the knowledge that real data were already arriving pervaded the conference.

I had lived through this transition from simulation to reality during my doctoral research in the early 1990s. I wrote an emulation[42] for part of the 'trigger' system built to select data from electron–proton collisions in ZEUS and spent two years checking that code to death. The idea

[41] Unusually, the conference proceedings became a highly cited paper. http://arxiv.org/abs/1012.5412.

[42] In FORTRAN 77.

had been to feed simulated data into both the real trigger[43] and my emulation, and check that we had identical results.

It had worked perfectly until we got our first real data. At that point the trigger *and* the emulation both began producing nonsense. It was not a good moment. All that careful preparation, and yet we couldn't make head nor tail of the real data. The only glimmer of hope was that the emulation and the trigger were producing identical nonsense, so I could use my code to investigate. After a few stressful hours, I had it. The wires in the detector were arranged in a pattern of little cells, each cell being a row of eight wires,[44] and each wire received an electrical pulse if a charged particle passed nearby. In the simulated data, the pulse had come out in order of wire number, 1–8. In the real data they came out in order of arrival time, which depended on where the particle was! Once we took it into account properly, all the crazy numbers lined up again. Obvious in retrospect, but somehow not anticipated. These things happen. We practise on simulations for so long, and it helps, but it can't fully prepare you. On the LHC, things were now very real, and nervous tension was everywhere.

3.3 Copenhagen

Within a couple of weeks, a steady stream of head-on proton–proton collisions was being delivered at a collision energy of 7 TeV (that is, two beams of 3.5 TeV each), allowing us to search for new particles, forces and dimensions in completely new territory.

In some optimistic scenarios, new particles or other exotica could simply have poured out of the collisions the moment they began. Like many physicists, I never quite believed that would happen, but we had to be prepared for the possibility. I suspected it was always going to be a long haul

---

[43] Which was in the Occam language and running on a network of transputers, if you care about such things. This was very cutting edge at the time.

[44] Numbered 1–8 rather than 0–7 because this was FORTRAN.

involving several years of hard, sometimes tedious, work. The LHC would be producing new physics for at least a decade. We weren't going to find the Higgs boson (if there was one) on day two, but the work had begun.

Even so, we had the big conference[45] coming up in Paris in July, and we were all hoping to get some preliminary, relatively simple results ready in time for that. In late June, we all headed off to our collaboration meeting in Copenhagen to prepare.

ATLAS is a big international collaboration. There are about 3000 of us, from 38 different countries. We have three big meetings a year, and more-or-less continuous smaller ones. We are the meeting on which the sun never sets. Of the three big annual meetings, two are typically in CERN while the other is in one of the collaborating institutes. That year it was Copenhagen's turn.

Much of the meeting was dedicated to results on how well the detector was working: did we understand what it was telling us? How good were the calibrations? Most of the rest of the time was for the discussion – and hopefully the approval – of preliminary results to be shown in Paris. The main result I was involved in was trying to understand and measure the jets[46] formed from quarks and gluons smashed out of the protons. We wanted to get a preliminary result written up in a 'conference note', which is the way ATLAS provides backup information about a result that has been shown in public but isn't quite ready to be submitted to a journal. (You'll hear more about these during the Higgs search.) This would have to be followed up by a paper as quickly as possible.

ATLAS, like most big collaborations, has a pretty intricate review process. It evolves, but basically goes something like this:

1. Have an idea for a measurement you would like to publish.
2. Do some work and present it in the relevant group. ATLAS has several physics groups, for 'Higgs' or 'Top Physics' or

[45] The International Conference on High Energy Physics, ICHEP. Like the LHC, a name designed for accuracy rather than inspiration.

[46] See Glossary: Quarks, Gluons and Jets (pp. 22–26).

'Standard Model', for example. You need to persuade the group, and particularly the group conveners, that your idea is a good one that ATLAS should publish.

3. They will request that an 'editorial board' be appointed. The ATLAS publications committee will appoint one.
4. Have meetings with your editorial board until they are happy with your analysis.
5. Present the analysis for approval in the physics group.
6. If it's not approved, go back two steps (this applies to most of these steps).
7. Now agree on the paper draft with the group leaders and the editorial board.
8. Distribute the draft to the whole of ATLAS. Wait for comments.
9. Present the result to ATLAS, addressing all the comments in a new draft.
10. Circulate the new draft to the whole of ATLAS again. Wait for comments.
11. Final presentation to ATLAS. Get approved.
12. Send draft for 'final sign-off' by some senior physicists on ATLAS (usually the spokesperson or deputies). If everything has gone well, this should be almost a formality. But it is the last chance for ATLAS to catch any mistakes. Also, if you are unlucky, you get a senior physicist who hasn't been paying attention to the first 11 steps, and wants to start all over again.
13. Send to journal, and to the arXiv (pronounced 'archive'. This is the place all good particle physics and astronomy papers are released, stored and freely accessible). The result is now public. Wait for comments from the referees.
14. Address referees' comments.
15. Journal acceptance. Now smile.

It is a lot of work, but generally it functions, as long as you have reasonable and diligent people involved in each step. At every step, confidence in the result increases, though mistakes do get found up to, including and unfortunately sometimes even after step 15. Journal publication is an important badge of quality for a scientific result, but it is not infallible. It will be useful to refer to these steps when discussing some of the rumours and leaks about the Higgs that emerged over the months that followed.

For a conference note, rather than a journal paper, steps 10 and 11 can be skipped, and 13, 14 and 15 don't happen. In Copenhagen many notes were passing steps 5 to 9. The note I was writing passed steps 5 to 8. Step 9 happened on 12 July, and step 12 went smoothly soon after. On 16 July, the jet cross-section note was released[47] in time for ICHEP, which started on 21 July.

I realise I haven't actually said what a cross section is. As it's an important and common concept in physics, it has its own section together with luminosity – see Glossary: Cross Sections and Luminosity (pp.72–4). Sufficient to say here that a 'jet cross section' is a measure of how likely we are to see hadronic jets when we fire two protons at each other.

When we collide protons, we really care most about the collisions between the proton's constituents – quarks and gluons. Unfortunately, the quarks and gluons only carry a fraction of the energy of the proton, and we have no way of choosing how much. If the fraction were a half, for example, then we would have jets with 1750 GeV of energy (half of 3.5 TeV). But most of the quarks and gluons carry much smaller fractions; our jet cross sections ran out of statistics at about 600 GeV. Still, since jets are the most commonly produced high-energy objects in proton collisions, this was a higher energy than anything else seen so far from the LHC, and it in fact had already reached the point where the Tevatron measurements had run out of steam. To have a real measurement of this, and to be able to show that the predictions of QCD agreed with the data, felt like a real achievement. Dozens of people were directly involved, and hundreds less directly. Like the minimum bias results, it was an important part of

[47] http://atlas.web.cern.ch/Atlas/GROUPS/PHYSICS/CONFNOTES/ATLAS-CONF-2010-050/

finding our feet in the new energy regime of the LHC. This is what we presented at the Copenhagen conference.

Despite this triumph, Copenhagen remains more famous for the role it played in the development of quantum mechanics, especially in the work of Niels Bohr and in the 'Copenhagen interpretation' of quantum mechanics. This is arguably the most flaky aspect of current physics, and unfortunately it's one that the LHC seems extremely unlikely to help us with.

In the Copenhagen interpretation, quantum mechanics only allows us to calculate the probability of something happening. All possible ways that the 'something' could occur have a number – an 'amplitude' – associated with them. You add up all the amplitudes, square the result and you get the probability. This is a very accurate way of making predictions if you have a large number of 'somethings'. In particular, amplitudes do not have to be positive numbers, so contributions can cancel each other out rather than build each other up, and this means that the probability of particles travelling from one place to another can exhibit interference effects.[48] Depending on the possible routes, there are regions of high probability and regions of zero probability, where everything cancels. This is behaviour we are used to seeing in waves, and the maths behind it is basically the resolution of wave–particle duality. Is an electron a wave or a particle? Well, it's a discrete quantum object, an excitation in a quantum field, a lot like a particle, but since we need to use amplitudes to calculate its behaviour, it can behave like a wave sometimes.

None of the above is the problem many of us have with the Copenhagen interpretation. That all works. The problem is, how do you move from a 'probability' that we have calculated to an actual occurrence. I can calculate the probability of an electron taking a particular route between an emitter and my detector, but what decides which route it will actually take? Even worse, what decides when the decision will be made? It doesn't bother me

---

[48] This is the physics at play in the WW scattering processes I discussed in 1.2. The 'No Lose' Theorem, for example.

so much that there is an indeterminacy in there, what bothers me is that at some point a particular outcome happens, and the transition between a superposition of amplitudes and a single result is not understood.

In the Copenhagen interpretation this decision point is called 'wave function collapse' and it is the point at which you put away the amplitudes and the quantum stuff and treat everything as definite. If you like, it's the point you open Schrödinger's box and give him the news about his cat. It is a very uncomfortable thing to have in physics because it seems to separate the observer from the observed, and we would like physics to explain both of them together as a single system. Cosmologists explicitly talk about the 'wave function of the universe', where the universe includes all physicists and their detectors, and their cats. How does that get collapsed, then?

It seems that as a quantum system gets bigger and more complex, the wave function does tend to collapse. Building large uncollapsed, or 'coherent', quantum systems is very difficult. There are rooms full of people in my department at UCL trying to do this, both to try and understand how it works and because if you can do it there are many applications, starting with quantum computing.

There are other interpretations, such as the many-worlds one, where all the possibilities are realised in a multiverse. (How did your consciousness end up in this one, then?) Several good physicists have spent time trying to develop hidden variables, but these make predictions that fail experimental tests, or at least the hidden variables must be very weird – they must be 'non-local', involving what Einstein disparagingly called '*spukhafte Fernwirkung*' or 'spooky action at a distance'.

I get invited to quite a lot of science-influenced art shows and exhibitions, of varying quality. Science is undeniably the source of some wonderful images. But speaking generally, the art that has most impact on me usually hints at, and shows back to me, something I have some knowledge of already, and leads me into a different way of thinking about it. This happens with art that is not specifically about science. It may refer to love, distance, location, parenthood, fear . . . almost anything.

This sets off all kinds of echoes in my thoughts and deepens the experience and understanding.

The only piece of art about science that has had this effect on me is related to this business of hidden variables and wave functions. It is *Copenhagen*, the play by Michael Frayn, which I saw years ago at the South Bank theatre in London and which still surfaces in my mind at random intervals, especially when I am working with or teaching quantum mechanics. Frayn brilliantly explores a meeting between Niels Bohr, his wife Margrethe, and Werner Heisenberg in Copenhagen during the Second World War. Bohr and Heisenberg, giants in the development of quantum mechanics, were long-standing colleagues, but on opposite sides during the war. No one really knows what was said, but afterwards the Bohrs fled Denmark to the UK and from there went to the USA (not a moment too soon). Heisenberg led German efforts to develop an atomic bomb. Bohr collaborated on the Los Alamos project.

Why exactly Heisenberg visited, whether he warned Bohr, whether key physics information, or misinformation, was exchanged, and what impact this had on the fact that Los Alamos succeeded and the Nazis failed, has been the subject of much speculation, and alternative versions are presented in the play.

All that is fascinating, but there is another layer to it. There is an unstated but to me very clear analogy between, on the one hand, known facts of history (the meeting definitely happened) and the unknown motivations and conversations that may have been behind them, and, on the other hand, the idea of measurable quantities in quantum physics (for example, the position at which an electron hits a screen) and the indeterminate states, the amplitudes, in between that lead to it. To calculate the final position of an electron you have to consider all possible paths it could take, and it makes no sense to say it took a particular one. This is more than just lack of knowledge, it is in principle unknowable. For me, this is reflected in the personality of Heisenberg, in that he probably had very mixed motivations and was probably not even clear to himself what they were, or which were dominant. So in a sense only the actions taken are real. The motivations are potentially unknowable and perhaps you have to consider all of them to understand the deeds.

Bohr left for safety, and Nazi Germany did not develop the atomic bomb. In happier times, the ATLAS collaboration left Copenhagen with a lot of new results to show in the big Paris conference.

Glossary: Cross Sections and Luminosity

'Cross section' is a common concept in particle physics, but a potentially confusing one. In this context it is basically a measurement of probability, except that probabilities have no units and are all between zero (never happens) and one (definitely happens), whereas a cross section has units of area and can take any (positive) value.

To see how this works and why, imagine trying to score a penalty in football without a goalkeeper present. If your skills are anything like mine (or if you are an England player in a shoot-out against Germany), there's a non-negligible chance you will miss. There are many factors to take into consideration, but the chance you will actually score will depend on the cross-sectional area of the goal. The bigger the area, the bigger the chance.

This (or equivalent experiments with nuclear beams fired at targets, such as Rutherford's experiment firing alpha particles at gold foil) is why when we calculate whether a particle interaction will take place, say between two protons at the LHC, we can express it as a cross section, with units of area.

If I were firing randomly at a wall with goalposts painted on it, the actual probability of me scoring with a given kick would be the area of the goal divided by the area of the wall, assuming I always at least hit the wall. If the goal is about 18m^2, and the wall is about 100m^2, and I scatter my shots evenly over the wall (OK, I'm not really that bad, see below), then the probability of any one shot going in would be 18/100 = 0.18. This is really a probability. It ranges from zero (goal of zero area!) to one (goal as big as the wall).

Particle and nuclear physicists measure cross sections in barns. A barn is just a unit of area. It was supposed to be something big and easy to hit (as in 'barn door') and is roughly the cross-sectional area of a uranium nucleus. It is small by everyday standards – 10^{-28} square metres, or one ten billion billion billionth of a square metre. This is a nonsense number, I know, and it gets worse. A femtobarn, which is the type of cross section we measure at the LHC, is a million billionth of a barn. We just have to deal with the fact that nature operates on a variety of distance scales and not all of them are easy to imagine, I guess.

As I hinted above, something else that affects the probability (of a scored goal, or a particle collision) is the number of attempts I make per unit area. In the penalty-shooting example above, if I scatter my shots across the wall equally with an area density of one per square metre, then I will probably score 18 goals. If I scatter them at two per metre, then most likely I'll score 36 goals.

The 'shots per square metre' number is the 'luminosity'. It has units of inverse area, or 'per square metre'. A particle collider like the LHC can increase the luminosity by either firing more particles (like me still shooting randomly but shooting twice as often) or firing them more accurately (like me shooting the same number of balls but concentrating on the $50m^2$ of wall nearest the goal, something of which I believe myself to be fully capable).

At the LHC we now measure luminosity in inverse nano- pico- or femtobarns. These are useful units for luminosity, because the cross section for a proton–proton scattering at the LHC to produce a Higgs, for example, is expected to be about 10,000 femtobarns. If you multiply femtobarns by inverse femtobarns (like multiplying the number of shots per square metre by the number of square metres for the football), you get the number of collision events. At the point that we had one inverse picobarn, or a thousandth of an inverse femtobarn, of luminosity, we might well have made ten Higgs bosons already.

Unfortunately, most of them would have decayed in ways impossible for us to distinguish from other kinds of events, so we would have missed them. We needed more.

3.4 Paris in the Summertime

And so to Paris. The world was waiting. Some of it, anyway – that fraction of the world with at least a passing interest in particle physics, which seemed to be a bigger fraction than we were used to. Even President Sarkozy was going to turn up. Not only would the first results from the LHC be shown, but the last major update from the Tevatron was expected (with more rumours than usual circulating that they might have seen something Higgs-like or otherwise surprising), and on top of all that there was action in the world of neutrinos and the usual potential for surprises from anywhere else in the field.

The ICHEP is split into two, with a series of small sessions running in parallel for the first three days on fairly focused topics, then a day off, followed by three days of plenary talks, where ideally the details are pulled together and we see how physics has moved on.

I registered, acquiring yet another geeky backpack, a pen and a map of Paris. Plus a dog tag with my name and 'ICHEP 2010' on it so that security wouldn't throw me out – more of a worry than usual given the President of the Republic's pending presence.

In one of the parallel talk sessions the audience were hearing how well the LHC detectors were working. I already knew that the answer from ATLAS was 'very well indeed', and the same turned out to be true for CMS, a major achievement and further evidence that neither of us had wasted the unwelcome extra year of preparation we'd had due to the failure in 2008.

Being a bit LHC-saturated, and partly for old times' sake, I went to another very crowded session where the latest HERA and Tevatron measurements of the W and Z bosons were being shown, along with the very first measurements of these particles from ATLAS and CMS. I got a curious feeling from this. To observe the W and Z bosons was by no means a surprise. They have been well measured in previous experiments. Rubbia and van der Meer won the Nobel Prize for their discovery at CERN in the 1980s. The LEP machine, which had preceded the LHC in its 27km tunnel, had made enormously precise measurements of the Z boson, and the Tevatron experiments had measured the W mass to better than one part in a thousand. Yet somehow, seeing them again, now in 7 TeV collisions, brought home to me in a surprisingly powerful way the anchor that experiments provide for our understanding.

The W and Z lead very fleeting lives, decaying almost immediately into other particles. They are gauge bosons generated by the symmetries of the Standard Model of physics,[49] and according to this model their masses come directly from the Higgs boson (or, more correctly, the field associated with it). In a sense they are partly made up of the Higgs. This is all very weird stuff. But there are also very definite predictions. For example, when you plot the distribution of electron–positron pairs in our collisions, you should see a big bump. And there it is. It works. Don't let anyone tell you quantum mechanics is only about uncertainty.

In fact, quantum mechanics, when used in conjunction with precision measurements of the properties of the Standard Model particles, allowed us to set some constraints on the mass of the Standard Model Higgs boson itself. This is because within the Standard Model, quantum corrections to some of the things we measure involve the Higgs boson indirectly,[50] and they depend on its mass. By this stage, the least well-known, most important parameters to measure in order to constrain the Higgs mass were the masses of the W boson and the top quark, both of which the Tevatron experiments were reporting on. By the end of ICHEP, these constraints told us that if the Standard Model Higgs boson existed, there

---

[49] See Glossary: Gauge Theories (pp.96–9).

[50] Via little loops in the Feynman diagrams, see Glossary: Feynman Diagrams (pp.145–50).

was a 95 per cent chance that its mass lay between 42 and 159 GeV. When the limits from the Higgs searches were included (from LEP and Tevatron) this became 114–157 GeV.[51] Of course these calculations were all made assuming that the Higgs existed and that the Standard Model was correct, so they were no substitute for actually finding the boson. But it illustrates again how predictive the Standard Model was, and how confident we were that we would either find the Higgs boson, or falsify the Standard Model, once we had enough data from the LHC.

Floating significantly more freely from the anchors of experiment in the hard-core theory session was Erik Verlinde of the Institute for Theoretical Physics in Amsterdam, discussing the idea that gravity and general relativity may not be fundamental, but instead may emerge from the bulk behaviour of smaller things. If I understand it right, this would make gravitational waves essentially no more fundamental than sound waves. Fascinating, possibly a new direction, but as the speaker himself said, the theory needs to make some experimentally testable predictions.

In the final session of the first day we had presentations on the Higgs searches at the Tevatron particle collider at Fermilab in Chicago. The room was packed and I ended up hanging around outside, peering through the doorway and chatting with Gavin and Mike Paterson from the *Colliding Particles* films. What we saw were the component parts of a number of different searches for the Higgs, carried out independently at the CDF (Collider Detector at Fermilab) and D0 (D-Zero) experiments. What became quickly obvious was that no one was going to announce a clear-cut observation of the Standard Model Higgs at this conference. What was not yet clear was how close we were, how much room the Higgs would have left to hide in (meaning what mass values would still be allowed for it), and whether there would be any hints of its presence. We'd have to wait for the combined results in the plenary sessions for that.

I loved Saturday's session on jet measurements. There were talks from

---

[51] These results were reported by Martin Goebel http://indico.cern.ch/contributionDisplay.py?confId=73513&contribId=313.

HERA, the Tevatron and the LHC, with the last including the jet cross sections I was involved with. Even though it switched the machine off in 2007, the HERA collaboration was still analysing its data and presented some really beautiful, precise measurements. The Tevatron also presented very good measurements. Of particular interest to me was its first measurement of the mass of a jet. Measuring jet mass is a good way to learn about QCD, but it is also important if you want to search for boosted particles using jet substructure, as I described in section 1.8. The CDF results showed that for 'normal' jets from quarks and gluons, the theory described the data quite well, to within a few per cent, which was good news.

One of the busiest sections of the conference was the session on neutrinos, an area of particle physics quite independent of the LHC that was making a lot of progress. I'll come back to what was happening there in later chapters.

As ICHEP moved from the parallel to the plenary sessions, some people left, some arrived, some, like me, stayed on. Sunday was a day off and featured the final stage of the Tour de France. The plenary sessions began with a grand opening by President Sarkozy, followed by the LHC summaries, a press conference, and the presentation of the combined, updated Higgs search results from the Tevatron by Ben Kilminster of Fermilab.

Doing the femtobarns/inverse femtobarns calculation for the Tevatron, it was possible to work out that if the Standard Model Higgs existed, about a thousand of them would have already been created at the Tevatron, with the exact number depending on the mass of the Higgs. But, as at the LHC, so at the Tevatron: they are incredibly difficult to dig out of the data. The CDF and D0 collaborations had now closed off a section of the possible Higgs mass range (158–175 GeV), leaving the boson less room to hide. This raised the stakes, both for the Standard Model and for supersymmetry (of which more soon). Both prefer lower Higgs masses than this, so ruling out high masses strengthens their case. But both need the Higgs to be there, so ruling out some mass range threatens them.

The statement was that if the Tevatron could run until about 2013, and if the analyses kept improving, it would be able to rule out the whole mass range for the Higgs at some level, or find a hint of its existence if it was there. This would have set up some kind of race over those years between the Tevatron and the LHC. It wasn't to work out quite that way.

Sarkozy's speech gave a strong endorsement to the place of fundamental science in our society and economy, and powerful statements on the need to make sure that current economic 'urgencies and emergencies' would not harm its long-term future. This was a refreshing change from the edgy situation in the UK, where the new coalition government was planning its first budget in the middle of the economic crisis, and the Royal Academy of Engineers had just released a policy document urging that 'The overriding consideration for BIS [the Department for Business, Innovation and Skills] should be the impact of research on the economy in the short to medium term,' with a heavy not-even-subtext that pulling out of CERN would be a great idea. Given the long-term symbiosis of great engineering and science at CERN (where the LHC was led by a Welshman and then a Northern Irishman), this felt like a massive betrayal, or as one might put it more diplomatically, a 'missed opportunity'. The battle over science funding would continue in the run-up to the budget. And for ever, I suspect.

The week of ICHEP was followed by a small add-on meeting at LAL Orsay, the French national physics laboratory to the south of Paris, so I ended up spending the best part of ten days in Paris (one day was spent in unlovely Swindon, back in England, at research council meetings, so I do mean the *best* part). I like Paris a lot, and Susanna came out for the weekend in the middle, so even better. My hotel was pretty dumpy but functional, and since it served breakfast in a nicotine-stained dungeon I went to a café every morning, where the waiters were friendly and the view was better.

The *Colliding Particles* crew (Mike and his soundman) were around most of the time filming, in part also for the *Guardian* newspaper. They had also been to the previous ICHEP in 2008 in Philadelphia. One of my colleagues had actually seen the Philadelphia film and thought it was made in Paris. Of course Philadelphia and Paris are not so similar, so it is disorientating when they suddenly show commonalities. We had a meal in a small bistro in Paris, apparently off the beaten track but it must have featured prominently in some key guidebook since it turned out to be full of American couples plus their children, all drinking water. I became so culturally disorientated I spilled my wine all over the table.

Enough tourism. ICHEP overall was a good reminder that older experiments can still throw up interesting stuff, even though new experiments may surpass them in many ways. The trade-off between old and new means the best physics from an experiment often comes close to the end of its lifetime.

All that said, the next two years would belong to the new kid on the block – the Large Hadron Collider.

3.5 Super Symmetry

I realise this is in danger of reading like a travelogue, but I sort of have to mention that a couple of weeks after ICHEP I gave a talk at the SUSY 2010 conference in Bonn, because it is about time to bring supersymmetry (SUSY) into the story. For a variety of reasons it is not possible to discuss LHC physics for long without talking about SUSY.

It is a bit weird that there have been *21* annual SUSY meetings, even though there is as yet no experimental evidence for SUSY playing any role in actual particle physics. Perhaps it's excusable. At least before the LHC switched on, SUSY was arguably the best way to improve the Standard Model of particle physics.

As you'll know if you read the Glossary on the Standard Model particles and forces, all the matter particles (quarks and leptons) are fermions, and all the forces are carried by bosons. You might (especially if you are a physicist) ask whether this is really a rule of nature, or a coincidence. What if you swapped all the bosons and fermions over, would the world be very different or not?

This is a very good question, by which I mean that asking questions similar to this has led us over the years to some very important and interesting answers. It is a symmetry question. Symmetry is probably the single most important concept in physics. One of the most important theorems we have, which applies to the classical and quantum regimes of physics, is Noether's theorem. This states (with some formal caveats that I am not going to go into here) that for every continuous symmetry in nature, there is a conservation law.

That may not leap out at you as being quite as profound and important as it is. But consider this:

1. Emmy Noether was a mathematician. She proved this. This is not just an opinion.
2. It means that because the laws of physics don't depend on where you are, momentum is conserved. Because the laws of physics don't depend on what angle you look at them, angular momentum is conserved. Because the laws of physics don't depend on when you look at them, energy is conserved.

That is a lot to get to grips with. But bear with me, it's worth it.

By the 'laws of physics' here, I mean the equations we use to describe the way physical systems behave. They are derived from a combination of observation and mathematics, they're provisional, this is not a religion (or a judiciary), and they work. When Noether talks about a continuous symmetry, she means, for example, there is a variable – call it x, why not? – in my equation meaning 'distance from my house', and if everywhere in the equation I replace 'x' by 'x plus y', where y can be any distance – a nanometre, a mile, the distance to the nearest chemist's – the equation still works. That's a continuous symmetry. It's called translational symmetry. If my equation (with x in it) is the equation describing the motion of some object, the equation really will look the same whatever value of y I choose. This symmetry leads directly to conservation of momentum, which is basically Newton's first law: An object will remain at rest or moving with a constant velocity unless acted on by an external force.

Step back and think about that (I'm doing so myself as I write this). Noether's theorem connects the idea that there is no privileged position in space – that the physical rules that the universe obeys don't care where you are – with Newton's first law. It's really pretty stunning.

Continuous, universal symmetries like that are very powerful. In fact, rotational and translational symmetries combined are part of a bigger symmetry called the Poincaré symmetry group, which seems to be the symmetry group of space and time. As well as translations and rotations, it includes the symmetry that physics is the same whatever velocity you are travelling with, which is basically the principle that underlies Einstein's special relativity. And if you think by bringing Einstein into this I am appealing to authority, I'm not. I'm just saying it works, and it's amazing.

There's another class of symmetry that is also quite amazing. This is the class of 'internal' symmetries, related to the quantum numbers a particle can have. A 'quantum number' is a property that is intrinsic to a particle, such as the charge of an electron, or the color charge of quarks. If you invert the charge of all charged particles at once – so all electrons become positively charged and all protons become negatively charged – the electromagnetic force will look just the same. You won't be able to tell the difference.[52] So 'charge inversion' is a symmetry of the electromagnetic force. These kinds of symmetries are also crucially important, and in fact all the forces of the Standard Model are derived from internal symmetries like these.[53]

So symmetry is a really vital element of physics. This applies to fundamental particles, but also to other areas of physics. It is one of the most powerful mathematical tools in our toolbox, and appears in the natural world all over the place. It is, in general, pretty super.

'Supersymmetry', or SUSY, is an extension of these ideas, and as I said before being diverted by Emmy Noether, it postulates a symmetry between bosons and fermions. In a perfectly supersymmetric universe, everything would look the same if you swapped bosons for fermions and vice versa.

---

[52] You will if you use the weak force, however.

[53] See Glossary: Gauge Theories (pp.96–9).

This is obviously not true – there is no boson around with the same mass as the electron, for instance – so supersymmetry can't be exact. But it could be a symmetry that is present in the underlying theory, but broken in everyday life. This could manifest itself as the fact that the bosonic partner of the electron (which we would call a selectron, for 'supersymmetric electron') has a mass much higher than the electron. But this mass might be low enough for selectrons (or other superpartners of Standard-Model particles) to be found at the LHC.

So while symmetry is firmly established as a useful principle in physics – and in particle physics in particular – supersymmetry has yet to prove itself. Why, then, have there been (at the time of writing) 21 conferences on the topic? As far as I can see there are three big arguments in its favour:

1. It helps with an important problem in the Standard Model.
2. It sort of predicts dark matter.
3. It looks nice.

The first of these is to do with the Higgs boson. Now, like SUSY, the Higgs boson had not shown up at this point. However, unlike SUSY, the Higgs boson is an integral part of the Standard Model, without which it doesn't work. There is a subtle problem with this, though. Because the Higgs boson, uniquely amongst all Standard Model particles, has no spin, its mass picks up a particular kind of quantum correction. If left alone to do their thing 'naturally', these quantum corrections tend to make the Higgs boson millions of times heavier than it has to be in the Standard Model. This was (and is) a real worry for the credibility of the theory. From one point of view, it makes the Standard Model look like a coincidence on the level of one in ten thousand million million (10^{16}). This is about *a hundred times less likely* than winning the lottery jackpot two weeks running if you buy a single ticket each week. SUSY gets around this because fermions give negative corrections and bosons give positive ones, so if there is an (even approximate) symmetry between the two, most of the corrections cancel each other out and the Higgs mass can be sensible without fine-tuning things to achieve such a crazy coincidence.

The second argument is to me the most compelling. Astronomical observations tell us there is probably some dark matter out there (or else we really do not understand gravity). Many SUSY models predict a particle that would be an ideal candidate for dark matter. It may be right behind you. When two different branches of science have problems that seem to converge on the same solution, look out for progress.

The third argument is essentially the fact that, as discussed, SUSY is a way of pushing ideas about symmetry, which have already been shown to be a great way of understanding nature, even further. Going back to the two types of symmetries (the Poincaré group of external, space–time symmetries, and the internal symmetries like charge), there is a theorem[54] that states that external and internal symmetries cannot mix up amongst each other. Internal symmetry operations turn one kind of particle into another (for example, the matter–antimatter symmetry operation turns electrons into positrons), whereas external symmetry operations move you around in space–time (for example, the translation symmetry operation just moves an electron to a different place). But swapping a boson for a fermion does both, because while it obviously turns one kind of particle into another, it also involves a space–time transformation, because spin is actually angular momentum. Angular momentum is a space–time property, conserved because of rotational symmetry. So SUSY is a special loophole in the theorem that says internal and external symmetries can't mix.[55] In fact it is the *only* such loophole in a four-dimensional theory like the one we need to describe our universe. Since all the other available symmetries are exploited in nature, with elegant and far-reaching consequences, it is very attractive to suppose this last available symmetry should appear too.

---

[54] Coleman and Mandula, *Physical Review* 159 (5): 1251–6 (1967).

[55] Which was therefore generalised to the Haag-Lopuszanski-Sohnius theorem, if you really want to know. *Nuclear Physics*, *B* 88: 257–74 (1975).

Those are three quite strong reasons for taking SUSY seriously. But they all have their weaknesses too. For the first one, maybe the universe just got lucky? Some string theorists might say we should be glad the odds were better than one in a 'landscape' of 10^{500}. Or maybe we're missing something subtle in the Standard Model that might force these cancellations, so they happen without that fine-tuning, a bit like cheating on the lottery. For the second of those reasons, well, there are other theories that can also produce dark-matter candidates. And for the third, we know that many beautifully symmetric mathematical ideas have wrecked themselves on the rocks of data. We shall have to wait and see.

More prosaically, and from the point of view of an experimentalist more practically, another important feature of SUSY is its flexibility. It can appear in many different guises in an experiment, to the extent that almost any weird event we see could (and will, I bet you) be interpreted as a 'hint of SUSY'.

For example, a big part of my doctoral thesis involved simulating a SUSY process that we might have seen at the electron–proton collider, HERA. When you whack protons and electrons together, one thing that might happen is that the quarks in the proton stick to the electron. This would be a 'leptoquark' (because electrons are leptons) and would be a sign of the unification of the strong, weak and electromagnetic interactions – so-called 'Grand Unification'. Very exciting stuff.

Just before we switched on, JoAnne Hewitt, a theorist at Stanford, realised that the signature of a leptoquark also looked like a particular form of SUSY. Herbi Dreiner, then a postdoc at Oxford (and later organiser of SUSY 2010), had realised that if so, there would be other ways it could decay, and he calculated them.

I remember his calculation being given to me on a napkin, but my memory may be embellishing here. Anyway, I wrote a program predicting how the events might show up in our detector, so we could search for them. Sadly, they never showed up, though we did have a bit of a false alarm at one point.

Of course, SUSY is one of the things we might find at the LHC. In fact

I have even written a couple of papers on some possible signatures. (Which is one reason I was talking at SUSY 2010.) Basically the 'boosted, jet substructure' type ideas I was trying to spread have some application to SUSY searches as well as to Higgs searches.

This flexibility makes SUSY a good test case for experimentalists to make sure we aren't missing anything. If we are alert to all possible SUSY processes, we are alert to a very wide range of weird stuff. However, when weird stuff doesn't show up, as so far it has not, that unfortunately does not disprove SUSY, it just rules out a given subset of SUSY models. This can be frustrating.

Still, to its credit, if there were no low-mass Higgs, SUSY would lose much of its attraction. It would not quite be ruled out, but it would certainly be relegated down the ranks of speculative theories. Conversely, therefore, if we were to find a low-mass Higgs, the search for SUSY would become much more compelling.

I can't really give an account of the meeting because I could only be there for one day. But even on that day you could see the change of mood now that people were finally showing real LHC data rather than simulations. So far the data hadn't broken new ground in the search for SUSY, but it was obvious it would not be long before they would. At the time of the meeting we had about three inverse picobarns of data.[56] After about 50 inverse picobarns we would pass the Tevatron in the search for SUSY – and within a year, we were expecting to have collected about 1000 inverse picobarns – that is, one inverse femtobarn. Not only were we accumulating a lot of data, we were doing so at an increasing pace.

3.6 Names, Fame and Citations

A citation is when one scientific publication refers to another. The number of citations a paper receives tells you something, but it is often quite hard

[56] See Glossary: Cross Sections and Luminosity (pp.72–4).

to work out exactly what. In experimental particle physics this is even trickier than some other fields, partly because the author lists are so long. The reasons they are so long are tied up with the challenges of how such a big collaboration actually works.

Counting how often your papers are cited is one way to estimate how much influence your work is having, and it's very tempting to keep watching them. Especially when assessments by funding bodies, or promotions, hang in the balance.

Before ATLAS started publishing, my most highly cited papers were measurements of proton structure from ZEUS. These were important measurements, and I helped build and run the experiment, but I made no direct input into some of those papers. ZEUS had about 400 physicists signing the papers. About 3000 people signed the first LHC papers.

This is common practice in particle physics, for good reasons. But people do ask how it is possible for individual scientists to make a reputation, even a career, in these circumstances. It is a good question. Funding agencies and interview panels ask it too. Mostly, the answer is that 3000 people is a big peer group. People in there will know who does what, who really contributes and who doesn't. References count, as does authorship of important internal notes often accompanying the collaboration publications. The ability to work hard and well in a collaborative environment is at a premium – this is part of what makes particle physicists employable outside of the area, according to friends in industry – and prima donnas and time-wasters get quite widely known as such.

Our author lists are long and alphabetical. We discuss trimming or ordering them every now and then, and have always concluded that to do so would at best waste even more time, and at worst break the experiment. We already have far too many meetings, rivalries and resentments, and arguing about author lists, or the ordering of author lists, would likely make this much worse. The current practice is not perfect, but it works, and we haven't thought of a better way yet. The author list is a tangible expression of the fact that we are, all 3000-odd (yes, odd) egos, in it together. We need that kind of solidarity when things go wrong. And they do go wrong.

The man who headed the UCL high-energy physics group when I was appointed to the faculty was Tegid Jones. He is a fount of wonderful stories (many involving opera singers) and responsible for several new particle-physics words in the Welsh dictionary. His stories include an account of the water-filled proton-decay detector he worked on in a salt mine – the IMB (Irvine–Michigan–Brookhaven) experiment. When first filled with water, the experiment leaked, dissolving quite a lot of the mine and almost drowning the PhD hopes of a future congressman, amongst others. On the next attempt, it somehow achieved a 490m siphon effect with a cesspit on the surface, which was very bad for the optical purity of the water. After the third filling, it worked, but they failed to find proton decay. Not due to incompetence, but because protons don't decay, it seems.

Sticking together under such duress requires that everyone is in it together, and for scientists that means the author list. In the end, as well as setting limits on the time taken for some proton decays at about 10^{23} times the age of the universe, IMB did observe neutrinos from Supernova 1987a. This event (in 1987, as the name suggests) was the first observation of neutrinos from outside the solar system. IMB also developed a lot of the techniques taken forward by Super-Kamiokande, the experiment in Japan that eventually made the first measurement of neutrino oscillations.[57]

With ZEUS and ATLAS, I'm lucky to have been involved in the start of two major particle-physics colliders. In the first year of my doctorate research on ZEUS, data-taking seemed a long, long way away. Life consisted of unintelligible code, meetings full of acronyms, and schedules that slipped by a month every few weeks. Like (I am told) war, the ZEUS UK software meetings consisted of long periods of boredom followed by sudden violent conflict. While some senior colleagues got all excited about obscure software, the ZEUS Central Tracking Detector (the main UK contribution to ZEUS and *essential for all the physics* – a phrase we all had tattooed on our foreheads to impress the grant panels) was leaking gas and looking like it might never work. HERA started late, and more

[57] See 5.5 Meanwhile in the Neutrino Sector.

slowly than planned, and the ZEUS tracker was missing a lot of electronics when we began. However, the leaks were fixed, the electronics arrived, and in the end the tracker worked brilliantly for 15 years. ZEUS (along with our rivals H1) produced the best information we have on the internal structure of the proton, and made major advances in our understanding of the strong interaction (QCD).

The LHC also had a very public failure at the start, of course. Throughout all these tribulations, the collaborations, and their unfeasibly long author lists, stuck together.

So, I have hundreds of scientific papers to my name, and my degree of direct involvement in them varies wildly. I have my favourites, of course, the papers in which I recognise my own words, plots and ideas, as well as the results of my experiment.

Even amongst these, the top two at the start of the LHC were funny ones. The top 'paper', which I actually partly wrote and edited, contains no real data and no original theoretical ideas, and is not even published in a journal. It's an 1852-page tome containing preparatory studies for using the ATLAS detector. It was useful (though now obsolete) and the fact that it was cited a lot showed the level of interest in ATLAS, so it's fair in that sense.

Next down was a real paper from ZEUS. We reported the mass we measured when two types of particles produced in our collisions were combined: neutral kaons and protons. The neutral kaon is (like the pion) a meson – that is, it consists of a quark and an antiquark bound together, in this case a mixture of strange quarks and down quarks.

We made this measurement because some other experiments had seen a bump in the mass distribution they got when they combined neutrons with charged kaons, which might have been the first observation of a hadron made of five quarks. In the Standard Model, all the hadrons we know of are made of either one quark and one antiquark (mesons) or three quarks (baryons). If it really was a five-quark thing – a pentaquark – this would mean:

1. Big physics news, and
2. There should be a similar bump in our mass distribution.

We indeed saw a bump, though it was not completely compelling statistically and not necessarily in exactly the right place. Anyway, we did our job, we reported what we saw, and this was during a flurry of excitement so we got cited a lot. Sadly it looks like that particular pentaquark thing was a false alarm. Our bump may have been real, but something else, less interesting. Anyway.

Further down were lots of papers I'm more pleased with, some of which I've already mentioned. But the point is to illustrate how dangerous citation counts can be as an indicator of merit. I'd happily lose the top two papers before most of the next ten, because the next ten contain more data or more original ideas. They advance knowledge more.

The idea that the planets orbit the Sun dates back at least to Aristarchus of ancient Greece. Even the work of Copernicus, who is credited now with the first accepted heliocentric model of the solar system, was ignored for many years. Copernicus and Aristarchus would have struggled for promotion and grants based on citation counts during their lifetimes.

It is sometimes hard to fathom why some good ideas or important measurements languish in obscurity for ages while others have a rapid impact. I often wonder how much the name of a project influences this. Maybe having a catchy, memorable name helps. But then what constitutes a good name? Particle-physics project names seem to come in two types: the fancy names and the simple acronyms. Sometimes (rarely) they are both.

The LHC is obviously in the second category. Large. Hadron. Collider. Does what it says on the tin, as long as you realise the collider is big, the hadrons are small, and you spell them correctly.

CMS too. Compact Muon Solenoid. Actually it's 21m long and 15m by 15m thick, but compared to ATLAS it's compact.

ATLAS is a fancy-name type, really, though it makes a stab at being an acronym with 'A Toroidal LHC ApparatuS'. Flaky, if you ask me. Just be out and proud with your big strong classical allusions, I say.

On HERA (Hadron Electron Ring Anlage. Nice – classical, catchy and a proper acronym) there was an experiment called H1. Logic demanded that the other would be H2, but I guess someone rebelled and called it ZEUS. All very erudite, if a little Freudian. It had a novel calorimeter, made of scintillators and depleted uranium borrowed from the US (we sent it back when we had finished with it). The wits of H1 decided ZEUS was an acronym for 'Zero Experience with Uranium Scintillator'. How we laughed. But the calorimeter worked very well, so who cares.

It does go to show that you can make almost any name into an acronym if you try hard enough. Even ALICE is A Large Ion Collider Experiment (not bad, actually). There are limits, though. One of my most often cited papers, which I do like, illustrates this.

Back in 1993, Jeff Forshaw and I put some cosmic-ray calculations together into a program to generate multiple quark and gluon scatters. We couldn't think of an acronym right off, but we needed to call our terrible FORTRAN 77 code something, so we operated under the working title of Jimmy Generator because it was mildly amusing and easy to say. We were confident we would eventually either work out what Jimmy could stand for or find a better name. The program lives on (after some help from Mike Seymour and a paper published with him in 1996). It was one of the main programs used to simulate the hadronic environment at the LHC during the first three years of running, and I have sat through many serious presentations with 'JIMMY' appearing on important plots. The paper with Mike and Jeff is now one of my highest-cited papers, and the program is still called Jimmy Generator. And I still don't know what it stands for.

3.7 Another Layer of the Onion

I wrote in section 2.6 about the calorimeters, which constitute one or two layers of the 'cylindrical onion' of detector technologies with which ATLAS and CMS surround the proton–proton collision point at the

LHC. Each layer tells us something distinct and important about the particles produced in the collisions, thus allowing us to work out the underlying physics.

The calorimeters measure the energy deposited by particles that stop in them – and by design that is as many particles as possible. The lead tungstate crystals and liquid argon of CMS and ATLAS respectively are chosen to be as dense as possible in order to achieve this. The lead crystals of CMS are wonderful objects. They are as clear as glass, but many times heavier.[58] They also, as an aside, absorb X-rays very efficiently: Dave Britton, who is now an ATLAS colleague but used to be at Imperial on CMS, told me of a time he took some crystals in his hand luggage from CERN to London for testing. They showed up in the scanner like lead bricks would. When his bag was searched, the security officer took an awful lot of convincing that the glass block he could see in the bag was the lead brick he had seen on the scanner.

But it was. Lead tungstate, like any good calorimeter material, is very dense. Hadrons, photons, electrons – all the particles stop. The two known exceptions are muons, which leave some energy behind but do not stop, and neutrinos, which leave no trace. Muons we can follow later, but neutrinos we can do nothing about, except deduce that they were probably there from the momentum imbalance in a collision.

However, even for the vast majority of particles, where the calorimeter can tell us their energy, we need more information. This is where the inner layer of tracking detectors comes in. They allow us get precise information on the actual path a particle takes from its creation in a collision to its impact at the calorimeter.

There are several reasons we need this information. For example, while it is usually a safe assumption that the particles came roughly from where the LHC proton beams cross, this is not a very tight constraint. We would

[58] At the CMS experiment visitor centre they have pairs of crystals, one glass, one lead tungstate. They look identical but you can pick them up and feel the difference in weight. Even when you know what to expect, it is remarkable.

really like to know very precisely – within a few tens of microns if possible – where they came from. This information can be combined with the information from the calorimeter to obtain a good measure of a particle's direction as well as its energy.

Also, there are usually several proton–proton collisions happening at the same time, a phenomenon we call 'pile-up'. This happens because the beams are dense, which is good for increasing the luminosity. However, pile-up is bad and confusing, since we really want to measure particles from an individual collision. If we can track a particle back to an individual collision vertex we can throw away the pile-up particles and focus on the ones we really want.

We can also see if a particle came directly from the collision vertex, or if some other particle was produced that travelled a little way and then decayed, giving a 'secondary vertex'. Tau leptons, and hadrons containing b quarks, generally do this, and detecting them is important for lots of measurements. For example, these are the two heaviest fermions the Standard Model Higgs boson could decay to (unless its mass was very high so that it could decay to top quarks). For some possible values of the Higgs mass, detecting taus and b quarks would be essential for a discovery.

Finally, if we apply a magnetic field, which we do, then the curvature of the path of a charged particle will allow us to measure its momentum. High-momentum particles will travel in a nearly straight line, while lower-momentum particles will bend a lot. The lowest momentum of all will go in circles.

So, for all those reasons and more, we build an inner onion layer of tracking detectors. The main technology used is silicon. Silicon is a semiconductor.

In an isolated atom, electrons are bound to the atomic nuclei. They are in discrete energy levels. We know this because we see electrons jump between different energy levels, emitting and absorbing photons with distinct energies as they do. Each photon's energy corresponds to the distance between two levels. In fact, by observing the emission and absorption of photons we can actually work out which atoms are present in a material. This is the field of spectroscopy, and it is how we know what stars are made of, even though we have never visited them.

These energy levels are the quantum-mechanical solutions behind the discrete electron orbits that Niels Bohr proposed in his model of the atom. This was the first model of an atom to have the nucleus and the electrons in essentially the correct relationship, and understanding it was a key point in the development of quantum mechanics as a theory.

Anyway, in an individual atom the electrons are stuck. What happens to this situation when you bring lots of atoms closer together to make a material?

In the case of an electrical insulator, nothing much. The electrons remain stuck to their individual atoms.

However, in some materials, metals for example, the upper electron energy levels of neighbouring atoms merge. This means that the electrons are free to move, without changing energy, throughout the material. Thus they can carry an electric current. This is an electrical conductor.

A semiconductor is, as you might guess, the borderline case. In fact, in its pure form, a semiconductor such as silicon is an insulator. Some energy levels have merged, but they have no electrons in them, so no current can flow. However, if there is a small impurity or fault in the material, a few electrons can escape into the merged energy levels and can then carry a current. By careful introduction of impurities, this effect can be controlled very precisely. This is the physics that lies behind the whole of the computing industry. Silicon chips are semiconductors with impurities that define exquisitely intricate circuitry, responsible for the words appearing on the screen as I type this sentence right now.

A semiconductor detector in particle physics exploits the same effect. When a charged particle passes through a semiconductor, it can knock into some of the electrons, giving them the tiny amount of energy they need in order to escape into the merged energy band. By applying voltages across the detector, we can make these electrons flow as a current, count them and work out where they were released. This tells us a particle passed close by, and from many of these 'hits' we can work out the path – track – of the particle.

This is the latest in a long line of particle-tracking technologies, and it is the best we have so far because it is very fast (important given the rapid collision rate at the LHC), very precise (we can tell to within a few tens of microns where a particle passed by) and also it takes very little energy to release an electron.

The last point is important because we want to measure the original energy of a particle produced in a collision. Every time it collides with a piece of material in our detector, its energy changes, and the original energy and direction are obscured a little. Every time it gives some energy to an electron and releases it, the original energy and direction of the particle are obscured further. And remember, this all happens before the particle reaches the calorimeter where its energy will be measured. With a semiconductor, the production of electrons is very efficient, so the amounts of material can be rather small and measurements of the original energy and momentum remain precise.

Even though these scatterings and energy losses are, hopefully, small, it is an important part of understanding and calibrating your detector to measure what they are. When constructing the detectors, we keep a record of how much material is in there and where it is – not just the semiconductor sensors themselves, but the data cables, the mechanical supporting structure, the high-voltage cables. The detectors also have to be kept cool (otherwise thermal energy can release electrons and give spurious signals) and so there is a whole infrastructure for doing this – in ATLAS we use C_3F_8 (octafluoropropane, a CFC) gas as a coolant.[59]

All the information from the construction is coded up into a computing model of the detector. There's an open-source software project called GEANT, initiated at CERN but now with many collaborators, which provides a toolkit for putting together materials and geometries and simulating how various particles interact with them and with electric

[59] We do not let it get anywhere near the ozone layer.

and magnetic fields. GEANT is now widely used in applications from space science to medicine. While we were using it to understand ATLAS events, Lewis Dartnell was down the corridor at UCL using it to see how deep underground on Mars any bugs would have to hide to escape cosmic radiation.

These semiconductor detectors only register particles that have electric charge. For example, neutrons and photons do not show up. However, photons are rather useful for working out whether our GEANT map of the detector is in fact accurate.

Occasionally a photon will interact with the material (if there is a lot of a material it definitely will, of course – this is what happens in the denser calorimeter). One of the things that can happen in this case is that the photon converts into an electron–positron pair. These particles are now charged and we can track them as they spiral away from each other. We can therefore see where they were produced. If you bathe the detector in enough photons (which the LHC certainly does!) then you can build up a map of the material from these production vertices, since the density of vertices will be proportional to the density of the material. The more material in a particular region, the higher the chance of an interaction, and so the more vertices.

This allows amazing maps to be made, where in the pattern of dots (one dot per vertex) you can see the silicon detector modules, the cooling pipes, the cables and the carbon-fibre supporting structures. We can make the same maps for our simulated detector, and check that we really do know what we are doing – and correct our simulation if we don't!

I presented all this stuff at a meeting at the Rutherford Appleton Laboratory in September 2010. The day ended with Jeff Forshaw arguing about what a particle *really* is, when you get right down to it. At 3 a.m. in the bar. I missed most of that, since I went to bed at 1 a.m., so possibly I still don't know.

But it's something to do with all those dots.

3.8 Into the Unknown

The first LHC paper really to probe for new physics was submitted by ATLAS to the preprint server[60] and to *Physical Review Letters* on 13 August 2010.

There had already been one paper (by CMS) on minimum bias results in 7 TeV collisions.[61] That was the first paper from the high-energy collisions. There were also numerous preliminary results from all the LHC experiments. But this ATLAS paper had got to step 13 of the approval process[62] and was therefore considered 'final' by ATLAS. It was the first paper to contain results from collisions between quarks and gluons at energies beyond the reach of the Tevatron. This was the point that we were really, confidently, reaching out into new territory.

The paper set exclusion limits. This means we hadn't seen anything unexpected, but we had pushed the boundaries of our knowledge of fundamental physics up a notch in energy. More data were still coming in and the measurements were getting more precise. The exploration had really begun in earnest.

Glossary: Gauge Theories

The concept of 'gauge symmetry' is at the heart of the Standard Model, and so I want to try and explain it. But to be honest, while I do understand the mathematics behind it, it is something I struggle to hold an intuitive picture of in my mind. Most of the things I talk about in this book I do have such a picture for – that's how I understand physics – but this one I don't, right now. So here I am trying to make it up as I go along.

---

[60] http://arxiv.org/abs/1008.2461.

[61] See 2.1 Low-Energy Collisions and Electronvolts.

[62] See 3.2 This Is Not a Drill.

Consider a snooker table (if you are American, consider a pool table). Consider the way the balls on the table interact with each other and how they move. They will obey Newtonian mechanics, trundling along at a constant speed in a constant direction until they bounce off another ball or a cushion. Eventually they will slow down and stop as they lose their energy of motion due to friction.

There is a symmetry within the physical equations describing how the balls behave. Actually there are several, but consider one of them. Imagine if the table is raised by 50cm. While this would make it harder to play snooker, it will not affect in the slightest how the balls move and interact with each other, as long as the table is raised by the same amount everywhere – that is, it stays flat and level.

This is an example of a 'global symmetry'. Some variable (the height of the table) changes everywhere at once (i.e. globally) and it has no observable effect on the physical system (the snooker balls).

Now, there are such symmetries in what we currently think are the underlying physical laws of the universe. I will give a real example shortly. But carry on with this case study for now. If the snooker table is the whole universe, it is a bit weird to think of it all changing at once. What, in fact, does 'at once' even mean? Since nothing travels faster than the speed of light, there is no absolute definition of everything happening 'at once'. Time depends upon speed, so what appears 'at once' to one observer will be a change that gradually moves across the table to another. The laws governing the interactions between the snooker balls should not depend on the speed of the observer, so this is all highly suspect. Put it another way – if the snooker table is 100 light years across and I raise one end of it, the soonest the far end can even know that the other end has been raised is 100 years later, since nothing travels faster than light.

So let's try making the symmetry bigger. Let's say the laws of physics have to be the same, have to be symmetric, under *local* changes in height, not just global ones. If the snooker table is being raised, to some observers it might all happen at once, but to others there will be moments when one part is higher than another. But for all of them, the physical laws must look the same. What are the consequences?

Well, obviously we will see some balls running downhill, or slowing down as they try to go uphill. If there is a dip in the table, balls will speed up as they run down it and collect at the bottom. If there is a hill, the balls will be repelled from it. These are observable differences, and they break the requirement that physics should look the same to all observers. To make the motion of the balls consistent for everyone, you have to introduce some kind of effective force acting on the snooker balls to get them over the hills and through the valleys – this is the gauge force.

This is not a perfect analogy, no analogy ever is. But the business of taking a global symmetry and making it local is what is called (for reasons that remain obscure to me) 'gauging' the symmetry. This really, really works and is very powerful. Here is a real-world example. It involves particles behaving like waves, so it's trickier than snooker balls, but it is no longer an analogy, it is a description of what actually happens in the theory, and the theory describes real life very, very accurately.

Electrons behave like waves. So they have peaks and troughs, and they have a phase that tells you when a peak (or a trough) is coming at you. If two electrons have peaks and troughs lined up, then the chance of finding an electron there is doubled. If the peak lines up with a trough, they will cancel out and you won't find any electrons.

The important fact here is that the only thing that makes any difference is the *relative* phase – are they lined up or not? If you change the phase of all electrons at the same time all over the universe by the same amount, absolutely nothing happens. This is like raising the snooker table while keeping it level and flat. It is another global symmetry. In group theory, this phase shift even has a name, it is the symmetry group called U(1). Actually, as I mentioned[63] when talking about Noether's theorem, this symmetry imposes a conservation law, which in this case turns out, remarkably, to be conservation of electric charge.

Like the snooker table, it is unrealistic, meaningless even, to think of changing the phase of every electron everywhere in the universe at the same time. So we should consider the possibility of it changing by different amounts in different places. And we should see what we need to do in order to make sure that the physical laws look the same even if this happens. Just as with the snooker table, what we have to do to achieve that is introduce a force. In fact, if you require that nature respects this U(1) gauge symmetry, you must introduce a very specific force: the electromagnetic force. In terms of quantum field theory, you have introduced a gauge boson – the photon.

Those words describe the maths; they aren't an analogy, they are a description in English of what happens in the equations. It is a beautiful thing to see.

There are other symmetry groups possible, not just U(1). Imposing a gauge symmetry based on the group called SU(2) gives rise to the W and Z. Using SU(3) gives you the gluon. This is why the photon, W, Z and gluon are called gauge bosons (and the Higgs boson is not, because uniquely it is not derived this way). This is why the Standard Model is referred to in the literature as a U(1) x SU(2) x SU(3) gauge theory. And this is why physicists sometimes seem obsessed with symmetry – it is beautiful, powerful, and it works.

The thing to take away from this, even if my pictures didn't work for you, is that all the forces in the Standard Model have to be based on these local – gauge – symmetries, and the bosons that carry them are therefore gauge bosons.

---

[63] In 3.5 Supersymmetry.

FOUR

Standard Model

October 2010–April 2011

4.1 Science is Vital

In October 2010 I went on a demo for the first time in more than 20 years. The previous occasion had been to shout either for the abolition of Margaret Thatcher's poll tax or for the retention of student grants. Averaging over both would give me a 50 per cent success rate. Probably there should have been others in between. Oh well. This one was for science.

To the extent that scientists form a community, it has rarely been mobilised in an overtly political way. Some individual scientists are very politically engaged, some are scientific advisers to the government, and certainly there have been lobby groups to protect research funding before. But I can't think of another time when such a broad-based coalition of scientists ever got together to voice a single message and hold a public demonstration in Whitehall, London, outside the Treasury.

The circumstances were unusual. The mobilisation for the Science is Vital demo was begun by Jenny Rohn, then a researcher in life sciences at UCL and a writer, along with CASE (the Campaign for Science and Engineering) and many other fellow travellers. Some of them (for example the former Liberal Democrat MP Evan Harris and Ben Goldacre of *Bad Science* fame) were longtime pro-science campaigners whom I had met during the libel-law campaign (still ongoing, though without demonstrations).

I was a participant rather than any kind of organiser, but from my point of view the message of the campaign was extremely well chosen.

We were in the middle of an economic crisis and we had a new government. Everyone knew cuts in government spending were on the way. The previous government had said lots of good things about science and under Lord Sainsbury as Minister of Science and Innovation these words had largely been matched by action. However, when he had stepped down in 2006 the wheels seemed to come off. In particular, massive cuts had hit particle physics and astronomy when two research councils were merged to form the Science and Technology Facilities Council, or STFC (presumably to the annoyance of neighbouring Swindon Town Football Club). Two subsequent science ministers had seemed either not to care or to be unable to do anything about it.

I had watched and participated in this long and very painful process as both a particle physicist and a member of various STFC committees. One high point had been a visit to see the Secretary of State responsible, John Denham, in July 2008, when I'd got to meet Peter Higgs for the first time. Peter Higgs was rather eloquent about the importance of fundamental physics, and John Denham and his then Science Minister, Ian Pearson, had listened and discussed it with us politely. But there had been many low points. In the end, after years of damage, Lord Drayson, the third science minister in the three years since the formation of the STFC, had set up a structure that, while it did not fix the damage that had been done already, did resolve some of the organisational issues that had contributed to the crisis.

After that nightmare, the particle-physics and astronomy communities were anxious to point out that they had already been severely cut, even before the financial crisis, and we were anxious that the good work done in the end by Lord Drayson should not be undone. The rest of the science community had not had to suffer all the problems of the STFC, but were also very worried that basic science might be seen as an unaffordable luxury in tough times rather than a critical investment for helping make times less tough.

The whole process had taught me that there are many different politically useful ways of making an argument. Winning an intellectual argument, or establishing the truth through evidence-based discussion, is one thing. Getting anyone to pay attention and act on the conclusions is another. Hence my presence at the demo. We had to tell the public and the politicians that the UK has something precious (as in valuable *and* vulnerable) in its scientific capability, and in fact in its research and education more generally. And it was not enough just to tell politicians behind closed doors, no matter how eminent the attendees at the meeting. The public needed to know and the politicians needed to know that the public knew. We could not afford to keep quiet.

When I was a PhD student, I did 'safety shift' on the ZEUS detector on one of its very first nights of data-taking (sometime in 1991). ZEUS was a massive particle detector, about 20m high and mostly hidden behind concrete shielding. Safety shift was a good one for inexperienced graduate students. Just plod around every hour reading dials and ticking a list, and report anything strange to the shift leader.

At some point during the shift, someone saw water dripping out of the bottom of the concrete shielding around ZEUS. This was very bad. A leak could do horrendous damage to the delicate instrument we'd spent years building.

People rushed around. The water was turned off, the procedure for opening the detector began, and various senior physicists appeared and went into a huddle with the shift leader.

Well below the level of this activity, I plodded on with my safety round. I noticed in the 'rucksack' (three floors of high-speed electronics in a metal box) that one or two of the temperature dials were slightly outside their allowed range. I went down to the control room again. Strictly speaking, I should report this. But everyone was so busy with important stuff. What to do?

For science funded via the STFC, the water had begun dripping in 2007. In the years between then and the Science is Vital demonstration in 2010, cuts in research grants of around 40 per cent had been imposed.

I had been sitting on various committees, trying to decide which great science to kill in order to try and save the rest. Stressful, unpleasant work in which the 'best' outcome is still dreadful.

There had been petitions, select committee reports and more. All through this, various important people in science policy were buttonholing scientists behind the scenes, saying things along the lines of 'Don't make a fuss, we see the problem and we'll sort it out. All this noise is counter-productive.' Sometimes some of us had believed them, not realising that often their only goal was to keep a lid on things while the policy was implemented. It's true that shouting, on its own, won't solve anything, and abuse is usually counterproductive. There need to be serious, sensible arguments based on strong evidence. But keeping quiet is a sure way to be ignored.

There had also been an undercurrent of 'Do you really want the public to know how much money we spend on stuff like astronomy and particle physics? Sure, *we* know it's not useless, but *they* won't understand, and if you make a fuss you'll get *no* support.' Thankfully, on that one we hadn't believed them. And at the demo we didn't just have our woe about cuts to tell the public. There was exciting science being done. The LHC was a big story, but there were plenty of others: the Planck satellite had been launched in 2009, for example, and there were great images from Cassini when it arrived at Saturn's moon. Somewhat to our surprise and relief, the public response had been overwhelmingly positive, even after the breakdown of the LHC in 2008. We also received a lot of support from fellow scientists who were sadly now in danger of being in the same boat and who were all represented at the demo.

Back at ZEUS, I nervously tapped the shift leader on the shoulder and showed him the reading. The effect was dramatic. He leapt out of the room, ran up the stairs and pressed the emergency power cut-off for the entire rucksack. They had turned off the cooling water but not the electronics. A few more minutes and the delicate, expensive electronics, the product of years of work, would have fried.

Carrying on doing science, if you are lucky enough still to be able to, can sometimes be the best way of influencing the outcome. But keeping quiet, no matter what the appearances, will get you nowhere and may be terminal. Scientists have to be part of the political debate about both their own funding and the relationship between science and the society that supports it and benefits from it.

When the budget settlement did come out, there was a cut for science, but less than most areas of government spending, and much less severe than some of the scenarios that had been proposed, which would have meant the loss of hundreds of research posts and studentships, and would most likely have forced us to close major scientific facilities in the UK or renege on some international commitments. Science is Vital was credited by some with being a significant factor in this relatively benign result. It is impossible to be certain how influential it really was, but it was our duty and we did it.

4.2 Science Board

Complaining about science-funding decisions is an occupational hazard of being a scientist. Rather than complain from the sidelines, I prefer to get involved (and then still complain). That's one of the reasons I spend quite a lot of my time on various research council committees, a job that can be both stressful and dull – a peculiarly unpleasant combination. On the other hand, you learn quite a lot about how things work, meet quite a lot of interesting and smart people, and get to see quite a lot of good science. Sometimes you are even able to fund it.

In October 2010 I was a member of the STFC's Science Board and we visited the Harwell campus in Oxfordshire. We saw the Diamond Light Source,[64] a bunch of really very big lasers, and ISIS. ISIS is a storage ring that provides beams of neutrons for lots of science applications. It is one of the things we were able to fund, at least partially, over this period.

---

[64] See 1.1 Why So Big?

Neutrons are hadrons; like protons they are made of three quarks, but because they have two downs and an up they have no electric charge.[65] The neutron was discovered by James Chadwick in 1932. Without it, atomic nuclei would not hold together. Neutrons can also break up nuclei – depending upon what one does with them this can be explosive, or can run a power station. Because they are neutral, you cannot use electric fields to steer or accelerate them. They do have a magnetic dipole moment, but it is very small, making it much more difficult to guide them with magnetic fields. At ISIS, a beam of protons is smashed into a target and the neutrons fly off – produced when the protons hit atomic nuclei in the target.

You can put various things in front of a beam of neutrons. Again, because they carry no electric charge, they ignore the cloud of electrons around atoms and molecules, and just 'see' the nuclei. How strongly they see them depends on the kind of nucleus. For instance, water molecules consist of two hydrogen atoms and an oxygen atom (H_2O), and neutrons scatter very strongly off the hydrogen – they can transfer energy to it very efficiently, largely because a hydrogen nucleus is simply a proton, which has almost the same mass as a neutron. (Just as in a head-on collision between snooker balls with the same mass, the incoming ball can stop dead, transferring all its energy to the ball that it hits.) This means that, studied with a beam of neutrons, water shows up very clearly. However, aluminium (along with many other metals) is almost transparent to neutrons (neutron-nucleus-scattering cross sections vary wildly with the mass, spin and internal structure of the nucleus). This was illustrated to us with a cute video of coffee being made *inside* an aluminium espresso percolator, seen through the walls by neutron scattering. More serious applications include watching liquid flow around inside a high-tech engine in order to study obstructions and optimise the design.

There are several other applications. Scientists and engineers come from all over the world to use neutrons from ISIS. Another major

---

[65] $-\frac{1}{3} - \frac{1}{3} + \frac{2}{3} = 0$.

commercial application is the investigation of the effect on electronics of the cosmic-ray air showers caused by high-energy particles from space hitting the atmosphere.

Electronic devices rely on semiconductors. Like the tracking detectors in ATLAS and CMS, the electrons in semiconductors need only a little nudge and they can carry an electrical current. This is why they are good for tracking particles, and it is also why we can use them to build complex solid-state devices by arranging exactly how and when any current will flow. Every electronic chip in the world is made like this, including some quite critical ones, such as those steering aeroplanes. Unfortunately, the electronics in these complex chips can still get a nudge from a passing particle. Not from the LHC, in general (though this does happen for the electronics we have near the collision point), but certainly from the particles in cosmic-ray air showers. This can turn a one to a zero in a computer's memory, which is not too bad if it just means your MP3 player skips a beat, but is terrible if the autopilot goes haywire.

Imagine a high-energy particle from outer space approaching the Earth. It will hit an atom in the upper atmosphere, very likely shattering it into hadrons, electrons and photons all travelling rather fast. They in turn will shatter other atoms and the fragments will shatter others, so that a shower of fast-moving particles develops, heading for the surface of the planet, with the number of particles growing as it gets lower. At some point – the shower maximum – the energy is shared out over so many particles that many of them no longer have enough energy to shatter atoms. At this point the number of particles in the shower starts to shrink as particles gradually slow and stop and do not set off any more showers of fragmented atoms. That shower maximum just happens to be about 10,000m on average, which is the typical cruising altitude for a passenger jet, meaning aeroplanes get a significantly heavier bombardment from cosmic rays than we do at ground level.

So you need to be *very sure* that any critical electronics controlling the plane can cope. The same applies to a lesser degree even at ground level, especially as electronic systems get smaller and faster. Putting your electronics in a neutron beam is a good way to test how likely they are to fail under cosmic-ray bombardments when they are at 10,000m. The neutrons fake the most dangerous cosmic rays.

I was particularly pleased to learn all this as in October 2010 I also became convener of the ATLAS Standard Model group, which meant a dramatic increase in air travel, essentially to the level of a weekly London–Geneva commute. We use teleconferencing and the web, but chats over coffees are still essential for proper coordination of what was by then a huge production machine for scientific papers.

Travel has always fascinated me. To be honest, it was part of the attraction of particle physics when I was applying for research studentships. At the time, I had never been on an aeroplane and my total experience of 'abroad' consisted of a couple of trips to Wales and one day in Normandy. Despite the fact that travel is now a regular part of my work, the fascination remains.

Having a routine, and a job to do at the other end of the journey, makes the whole thing different from holiday travel. Conference travel is not routine – often the place is new and exciting. But big labs such as CERN (Geneva), DESY (Hamburg), KEK (Tsukuba, near Tokyo) and Fermilab (Chicago) just become another place of work. I'm glad I have never had a regular long-haul commute like many particle physicists do. At least when I read my kids bedtime stories by teleconference, my evenings coincide with theirs.

A strange feature of routine travel is the way disconnected bits of the world become as familiar as your street back home, but are separated by huge expanses of the unknown. This seems weird to me and reminds me of the way I treated the Tube map when I first moved to London: I'd pop out of a station and have no idea how to get to another except by going back down underground again. When I finally merged my Kentish Town and Trafalgar Square islands of knowledge, via Camden Town and Bloomsbury, I felt a strange sense of security that I hadn't even realised I was missing before.

I'll never walk from London to Geneva, not even for charity. Commuting over such distances is only made possible by technology. Technology also helps us cope with the disconnects though, especially the Internet and the social networking it carries. Perhaps this applies to islands of cultural and social knowledge as well. There are people I know well, whom I trust and would go to for help if needed and a drink if not, scattered all over the world. I often wonder whether the combination of travel and remote two-way communication – person-to-person via public communication like social media, rather than broadcast media – is eventually going to make a real global civilisation. By that I mean one where most people's social network is way more geographically diffuse than ever before. The decisions we make, or which are made on our behalf by companies and governments, have been having global consequences for many years. Powerful people have been able to broadcast their points of view globally for ages. The fact that two-way, small-scale relationships are also possibly becoming global and routine represents a big opportunity to correct an imbalance. I wonder what we'll do with it?

4.3 Prospecting and Surveying

In those first couple of years of LHC data-taking, the Standard Model group on ATLAS was responsible for making many of the first measurements. Everything from the average number of particles produced in a collision,[66] to some of the rarest events, such as those in which two W or Z bosons were produced. By this stage we had millions of 'minimum bias' events, and only a few dozen containing a pair of Z bosons.

'Standard Model group' is a bit of a misnomer really. Some bits of the Standard Model – the top quark, b-quark decays, and most notably the search for the Higgs boson – had their own separate groups. Also, calling it the Standard Model group sort of implies that we already knew the answer – that we would measure the Standard Model. Of course we

66 See 2.2 Minimum Bias.

didn't really know that, at least not for everything we were measuring. Some things were being measured for the first time, and all of it was being measured at a higher energy than ever before. The key thing was that we were in general measuring processes for which the Standard Model made predictions. Often the theory made rather precise predictions, so it was a challenge to make precision measurements to compare to them.

In all cases, agreement would be a confirmation that the Standard Model worked in a new process, and a disagreement would mean either that the Standard Model was wrong or that there was a mistake in the calculation or (perish the thought) in our measurement. Convening this group as the first LHC data came in was essentially my dream job: helping ATLAS digest, understand and publish the vast amount of first-time physics information buried in the millions of proton–proton collisions the Large Hadron Collider was giving us.

The publications from ATLAS could be pretty much divided into two types. There are the searches for new stuff, and there are the measurements of new stuff.

The first type is made up of the 'prospecting' papers. Prospectors zoom into the new landscape of physics to which the LHC has given us access and look for quick wins, for really unusual and surprising features. If they don't see them, we have learned something about the new land – 'Hey, no gold volcanoes yet!' (Or no supersymmetry yet, or no black holes yet . . .). Of course, if anyone finds a gold volcano, they've got it made. And those things seriously might be out there; similar surprises have turned up before, many times. We were only in the foothills so far.

Sort of alongside but often slightly behind the prospectors come the surveyors. They study the new land, measure it, see whether it really complies with our best understanding of geology and suchlike. If it does – 'Success!' We have extended the validity of our current theories. If it doesn't – 'Success!' We have found what might be a gold volcano hidden under something else, and we and the prospectors can dig a deep mine together to get at it. Examples include measuring jets, isolated photons or W bosons produced in the highest-energy quark and gluon collisions ever seen.

There was a third type of result on the way, of course. This would come from the Higgs group, and from that part of the landscape where our understanding of the way things work meant that there had to be a gold volcano (OK, a Higgs boson in this case); otherwise the Standard Model would have failed. Those searches were starting in earnest around this time. Nothing was published by this stage, but there were enough collisions collected that staring at the plots and trying to guess if there was a Higgs boson hidden in them was starting to become a pastime, for me as much as for everyone else at the LHC.

When it runs, the LHC runs around the clock, seven days a week. The pre-accelerators accumulate and accelerate enough protons to fill the LHC, the LHC then accelerates them the last bit (from 450 GeV up to 3500 GeV at that time) and will then store the two counter-circulating beams for several hours, some of the protons colliding on each revolution. The shift crews on the experiments struggle to keep the detectors recording data, and to keep our eyes open. Every time the LHC is filled with protons, more landscape is opened up. The pressure was on to study it quickly, but even though it's a frontier, no one wants to be a cowboy. It has got to be done right. If we get it wrong, no one would die, and it wouldn't destroy the world, but it would waste time, could send us down blind alleys and, since the data would continue to flow, the truth would come out in the end and mistakes would be found out. That could range from mildly embarrassing to career-endingly awful, depending on the circumstances. Even the prospectors have to be careful.

4.4 Antarctic Interlude

Away from the LHC, other physics was going on. Ryan Nichol, who works upstairs from me at UCL, gets together with NASA every now and then and flies a balloon around the Antarctic. He even gets to visit the continent occasionally, which is both better and worse than Geneva. The balloon is very large, bigger than Wembley Stadium which seats 90,000, when fully inflated, and it carries an experiment called ANITA (Antarctic Impulsive Transient Antenna. An experiment name that is almost, apart from the appropriated N, both fancy and an acronym).

ANITA is looking for clues to one of the great questions in astrophysics: Where do cosmic rays come from, and how are they produced? We know that there are really, really high-energy particles hitting the Earth (and sometimes aeroplanes) all the time from outer space. The energy spectrum of these particles extends up to enormously high energies – more than 10^{20} electronvolts. Remember the LHC beams are at a few thousand GeV, or a few trillion (10^{12}) eV. The highest-energy cosmic-ray particles have energies a billion times higher than that! Imagine what kind of accelerator produces those. Actually, people have tried to imagine. Spinning neutron stars, black holes in the centre of galaxies, and supernova shock waves are some of the more prosaic proposals. Decays of super-heavy dark-matter particles and 'topological defects' – boundaries between bits of the universe that cooled down differently after the big bang and could manifest themselves as cosmic strings or magnetic monopoles – are some of the more exotic. In general, the relative number of neutrinos and other particles produced will be different depending on which, if any, of these models operate.

There is another source of neutrinos. The universe is filled with a bath of very low-energy photons left over from the big bang – the cosmic microwave background. Protons travelling through the universe will interact with these photons, and if their energy is high enough they can collide with these very low-energy photons and produce a new kind of particle (called a Delta – a particle like the proton but with a higher mass). Because this becomes possible, the probability of the collision goes up, and that means the cosmic-ray protons at high energy are attenuated. However, the Deltas will decay producing pions, and the pions then decay producing neutrinos.[67]

[67] So-called GZK neutrinos: Kenneth Greisen, Georgiy Zatsepin and Vadim Kuzmin (1966).

It is also true that there could be some very interesting physics going on in the collisions between these particles and the atoms in the atmosphere. The fact that they hit the stationary atmosphere, rather than another particle with just the same energy coming the other way, means the amount of energy available to make new particles is reduced, but it is still 100 times more than the LHC. Again, the presence of such collisions throughout the universe is one of the main reasons we knew the LHC would not cause a catastrophe.

We would really like to know where they are coming from, and what kind of extreme conditions produce them. The aim of ANITA is to address this by looking for neutrinos. This might not seem an obvious choice, since neutrinos are famously hard to find. But that same property means they are not affected by all the material and magnetic fields in the universe between wherever they started and ANITA, so they should, if we can measure their direction, point straight back to the source.

What ANITA actually detects is short radio bursts, using them to build up an image of the continent. Another – along with coffee seen by neutrons – in a series of weird ways to look at the world, I guess. They can measure the polarisation of the radio waves – which direction the electromagnetic field oscillates in as the waves travel. Neutrino interactions produce vertically polarised pulses. In the first ANITA flight, though, they didn't find any of these, and so, no neutrinos. But they found something else. They saw 16 pulses of horizontally polarised radio waves.

These turned out to be the signature of cosmic-ray air showers. In these showers, electron–positron pairs are produced and they spiral around the Earth's magnetic field lines, giving a characteristic radio signal seen by ANITA. As Ryan put it:

> The cosmic-ray air-shower radio signals were really unexpected and we only found them by checking a 'background' event sample for the neutrino search. It took us a long time to understand their significance to the point that for the second ANITA we removed the horizontal polarisation from the trigger to maximise neutrino

efficiency. Whoops! Needless to say we will be reinstating it for the third flight.

These showers have been seen before, for example by the Auger experiment in Argentina. But new ways of seeing them, and measuring where they come from, are valuable. ANITA had, serendipitously, demonstrated an important new technique with a lot of potential.

Neutrinos from the Sun, and those produced in cosmic-ray air showers, reactors and accelerators, have taught us a lot about particle physics. The hunting for neutrinos from beyond the solar system could tell us a lot, too, about particle and astrophysics. As well as ANITA, the giant IceCube array was looking for them. IceCube is a cubic kilometre of Antarctic ice instrumented with photomultiplier tubes to detect the light produced when neutrinos interact with the results. It would see the first high-energy neutrinos three years later.

4.5 Inside a Proton

Most of the time (though not all), the LHC collides protons, which are not fundamental but are made of quarks, bound together by gluons. When doing physics with proton collisions, it is important to know this, and to know as much as you can about how the quarks and gluons are distributed inside the proton.

At some level a proton is a nuclear family of two up quarks and a down quark, but if you look inside you see all kinds of mess. First of all, the quarks are exchanging gluons between each other, which is why they are bound together inside a proton. But even looking (or trying to look) at a single quark, surprising physics emerges.

If the proton were made of homogeneous mush, the smaller the wavelength of photon you use to probe it, the less of the mush you would see. But if you are probing quarks inside protons, once you have enough resolution to see them, you would expect things to look pretty much the same no matter how much you increase the resolution. This is because if quarks are fundamental, it doesn't make much difference how closely you look, they look like tiny points. This prediction is called 'scaling'. It was observed in experiments at SLAC at the end of the 1960s, and was compelling evidence that the quarks Murray Gell-Man had postulated to explain the patterns of hadron masses and quantum numbers were actual physics objects. The observation of scaling won the Nobel Prize in 1990 for Jerome Friedman, Henry Kendall and Richard Taylor.

But theorists noticed that if quarks can radiate gluons, the scaling doesn't quite hold. And according to QCD, they are rather likely to do this. If you try to observe a quark, for example by hitting it with a photon, you are likely to see it after it has undergone some of these radiations. Since there are three quarks inside a proton, you might think that on average they would be observed to carry about a third of the proton's momentum each. However, the fractions (usually called, imaginatively enough, x) that we see them carrying are typically much less than this, because radiated gluons carry the rest.

If we are looking at a quark by hitting it with a photon, then the wavelength of the photon sets the resolution – basically, how closely are we looking? Photons with a short wavelength can see shorter distances, and so can tell whether a quark has radiated a gluon or not, even if the gluon is still very close to the quark. Longer wavelength photons will not be able to resolve such a gluon from the quark and will see what looks like a single quark, with a momentum that is the sum of the quark and the radiated gluon. Short wavelength corresponds to high momentum,[68] and so high-momentum photons will see more quarks in the proton, at lower x, than will photons of lower energy.

There is an awful lot of physics in this. It's one of those stories of more precision revealing more. To put it the other way round, the closer you look at a quark, the more of the gluon radiations you can resolve, and therefore the lower the momentum fraction, x, carried by the quark, that

---

[68] We'll return for more on this in 7.3 Waves.

you see. This violates scaling, and the amount by which scaling doesn't quite hold can be calculated in QCD. These 'scaling violations' also agree with what is now much more precise data, mainly from the experiments at HERA. This is one of the cornerstones of evidence that QCD is the right theory for the strong interaction.

It is fairly amazing that such a complex object as the proton, with all of this going on inside it, will, left on its own, last practically for ever without falling apart. 'For ever' isn't something we can measure, but we do know that they last for at least 10^{29} or so years on average. We know this because experiments like IMB,[69] and now Super-Kamiokande, have watched lots and lots of protons very, very carefully for a very long time, and none of them decayed. Given the universe seems to be about 1.4×10^{10} years old, the limits set on proton decay mean the current limit on the lifetime of your average proton is 7,000,000,000,000, 000,000 longer than the age of the universe.

Apart from being a mess of quarks, a proton is also a hydrogen ion. Hydrogen, the most common element in the universe, is just a proton with an electron stuck to it. Along with helium, most hydrogen was made very shortly after the big bang, and everything else came from fusing these two elements together in stars much later. In this process, some protons are transformed into neutrons. However, the vast majority of protons in hydrogen have been that way for about 13.8 billion years.

So, left to themselves protons are OK. Of course, we don't leave them alone, and we smash some of them up in the LHC. Smashing things up is a time-honoured particle-physicist method of finding out about them. When I worked in Hamburg on ZEUS, we were smashing up protons with a beam of electrons. Smashing stuff works sometimes. A friend of mine bought a melon from the Fischmarkt on a Sunday morning (after a long Saturday night) and in the taxi home became convinced that he had bought a pumpkin by mistake. It was only when he threw it away in disgust and it smashed on the pavement that it was confirmed as an actual melon.

---

[69] See 3.6 Names, Fame and Citations.

One thing we can learn from smashing protons at HERA, and the LHC, is how the quarks are distributed inside the protons. For example at the LHC when W bosons are produced, a positively charged W (W^+) can be made by an up quark annihilating with an anti-down quark, and a negatively charged W (W^-) can be made by an anti-up quark annihilating with a down quark. The anti-up and anti-down quarks are part of the scaling-violation business in the proton. Not only can the quarks radiate gluons, those gluons can split into quark–antiquark pairs. In other words, a proton contains gluons, but it also contains many more than three quarks, and lots of antiquarks too. But if you cancel every antiquark off against a quark, you are still left with a net total of three quarks. And since of these there are twice as many ups as downs, the relative rates and distributions of W^+ and W^- production give information on where these quarks usually are inside the proton. Towards the end of 2010 the ATLAS and CMS experiments both published measurements of this, followed up by increasingly precise measurements later. Combined with data from HERA and other experiments, this improved our knowledge of the internal content and workings of the amazing proton.

I find it extraordinary how rich the phenomenology of QCD is. The equations for QCD (what we would call the Lagrangian,[70] the bit that gets written on souvenir mugs and T-shirts for the CERN gift shop) fit on one line, and to me at least give no real clue that things like hadron masses, and scaling violations, are implicit in their structure. Many people have spent their careers studying, calculating and measuring the implications of this equation.

This – studying QCD – was what I was doing when I was working at HERA. This is not what most people assumed I was doing when they heard I was working in Hamburg. Usually the first thing they would say was, 'Are you in the army, then?' (Presumably this would be in some role

---

[70] After Joseph-Louis Lagrange, or Guiseppe Lodovico Lagrangia as he was born. It is a function that describes a physical system. Actually, it is just the kinetic energy minus the potential energy, but it still manages to be way too convoluted to show here.

not involving haircuts or heavy lifting.) The second thing they were likely to say was something like, 'Reeperbahn, eh? Phwoar!'

I did like the Reeperbahn, the city's red-light district. It is sort of a more explicit version of Blackpool, really. Actually I haven't been to Blackpool for ages, maybe it too is explicit now. When I had visitors, an evening on the Reeperbahn was pretty much obligatory. These evenings kind of blur into each other, except for one particularly bizarre all-nighter that culminated in the most surreal experience.

Usually we'd sit in various pubs, then go dancing. The sex-industry stuff was really only part of the scene; there was some actual good nightlife there too. However, one night three of us decided to thoroughly visit some of the strip bars and pole-dancing dives. (Not the brothels, though.)

We chose carefully, making sure of a 'free' drink with the entry fee so that we didn't lose out even if we had to leave in a hurry. Some of the dives were enjoyable, but the last one wasn't. It was about 4:30 on Sunday morning by then, it had been a long night and I expect no one involved really wanted to be there. This was a large number of 'free' drinks later, so I wasn't looking around very carefully when a rather large, naked foot appeared way too close to our noses. I put the almost full bottle of lager in the inside pocket of my jacket, and we left in a hurry.

One great thing about a Saturday night in Hamburg is the Fischmarkt (the source of the melon/pumpkin confusion earlier) that follows on Sunday morning. Opening at about 5 a.m., it's an invigorating mix of very proper Sunday-morning shoppers and the dazed-and-confused remnants of the night before. Like us. There are fish, plants, bad bands, melons, coffee, beer and, most importantly, Bratkartoffeln mit Spiegelei – an enormous mound of fried potatoes with three fried eggs on top. Just the ticket.

Annoyingly, when I got to our table, having negotiated my way unsteadily across the cobbles from the big pan with the potatoes, I was one egg down. Looking back, I saw it lying on the cobbles, glistening, still whole and unsullied, yolk-side up. Of course, I went back for it. Unfortunately, as I bent down to get it, some git poured beer on the floor right next to it. I stood up and looked around, but they'd gone. I bent down again. More beer!

Eventually I gave up and made my way with my two remaining eggs back to my by now hysterical companions, who had been watching the whole thing. I complained to them about the beer, and when they could speak, they explained that the bottle in my jacket pocket had been responsible.

This wasn't the surreal thing.

That came about eight hours later when, after an eel curry and not enough sleep, I went swimming. (I was a lot younger then.) Having swum, I was sitting beside the pool, drinking hair of the dog in the civilised way you can in Germany, when the lights suddenly dimmed, everyone got out of the pool and some strange music started. A line of small children, aged between five and ten and wearing odd hats, trooped in. They climbed into the pool one by one, and as they did so, an adult lit the candle that was mounted right on top of each child's head. They swam round the pool (there were about 40 of them, if I remember right), the little ones barely keeping the candles above water. Then they climbed out. As they did so, the adult snuffed out each candle in turn. The children trooped out, the music stopped, the lights came on, and everyone carried on as though nothing had happened. I still have no explanation. I blame the eels.

I digress. Back in 2010, Christmas was coming, and the LHC had a new treat in store for us.

4.6 Heavy Ions for Christmas

The Large Hadron Collider is called that because it is large and it collides hadrons. Up until November 2010 it had been colliding one type of hadron – protons – so it might as well have been called the Large Proton Collider. At the end of the 2010 run, however, it showed its versatility. Lead ions, containing both protons and neutrons, were loaded into the tunnel and smashed into each other. The LHC became a heavy ion collider for a few days.

Although the beam energies for the lead nuclei were much higher than the proton-beam energies (575 TeV compared to 7 TeV) the energy per nucleon (nucleon being a generic term for protons and neutrons) was lower, only about 1.43 TeV. Thus the average energy per quark or per gluon was even lower. When it runs in this mode, the LHC is not really an 'energy frontier' machine.

Even so, you can legitimately describe both the proton and heavy-ion physics programmes as 'probing the first moments of the big bang', and it is worth examining why that is the case, and why they are different.

If you start from the present day and observe that the universe is expanding, then it is reasonable to suppose that in the past it was smaller. Now it is tempting[71] to reason along the lines that, since energy is conserved, this smaller universe contained the same energy as the present universe, but in a smaller space. Thus the energy density, which is essentially the temperature, was higher. There is a potential flaw in this argument, however. Remember that Noether's theorem[72] connects conservation of energy with the fact that the laws of physics do not change with time, effectively saying that all times are the same as far as physics is concerned. However, if we are considering cosmology, clearly all times are not the same. There is now a zero time – the big bang – which all observers agree on. We can measure absolute times relative to this; that's what we call the age of the universe! This implies that physics does not necessarily have to look the same – all times are not the same, and there is a universally agreed 'zero time'. So can we really rely on energy conservation when we are talking about the whole universe?

In a sense, yes. The master equations of cosmology, in general relativity, do not change, so they remain the same even if you choose a different origin. Therefore there is some definition of energy you could take that

---

[71] And I've done it myself sometimes. Sorry.

[72] See 3.5 Super Symmetry.

would be conserved. But this would have to include the energy stored in the gravitational field, which in general relativity means the curvature of space–time. Some cosmologists choose to do this and rescue conservation of energy, and some do not. Either way, the physics remains the same – it is just a matter of different interpretations of the same equations.

It turns out that when doing the cosmology properly, the main conclusion is the same. The average temperature of the early universe was higher than the current average. This is because the early universe was dominated by matter and, even earlier, by photons. This in turn means that all the particles in it were, on average, moving faster, and therefore colliding with each other at high energies – energies that increase as you approach the big bang from the present day.

As particles collide at higher and higher energies, different physical effects occur. For example, if atoms collide with high enough energy, they knock electrons off each other – they ionise. While the temperature was high enough for this to happen often, therefore, the universe was filled with plasma – a mix of ionised atoms and electrons. Light cannot travel through this; it keeps getting scattered by all the charged particles.

At some point the universe cooled to the extent that the typical collision energy was too low for ionisation. This meant that atoms and molecules with no net electric charge could form, and stay together. In turn, this meant that light could travel much more easily, since the photons were not continually interacting with charged particles. These photons, which were all over the universe at this point, are still travelling. They are much cooler now – about 2.7 kelvins, or -270°C – and they were spotted in the 1960s by Penzias and Wilson (who initially mistook them for bird poo).

Experiments mapping the cosmic microwave background, such as COBE (Cosmic Background Explorer), WMAP (Wilkinson Microwave Anisotropy Probe) and most recently the Planck satellite, look at the physics from this moment, about 400,000 years after the big bang, when the first atoms formed. From the fluctuations and frequencies of these photons it is possible to get clues as to what happened before, during those 400,000 years. But the collision energies at the LHC allow us to directly study the physical processes that must have dominated the universe then.

A few minutes after the big bang, the collisions were so violent that even atomic nuclei could not hold together. At this point, protons and neutrons were everywhere. These are the kinds of energies you need for nuclear fusion, as is being attempted at the International Thermonuclear Experimental Reactor (ITER).

Back a big step further (to about a millionth of a second after the big bang) and the protons and neutrons can't even stay whole. The quarks and gluons that they are made of spread over the whole universe (which is quite small at this point). This is a new form of matter we refer to as 'quark–gluon plasma', though experiments at the Relativistic Heavy Ion Collider (RHIC) at Brookhaven National Laboratory indicate it may actually behave more like a quark–gluon liquid. This is the stuff that the LHC could reproduce in November 2010 by colliding lead nuclei. (RHIC used gold nuclei rather than lead. It's flashier, but makes no difference to the physics.) ATLAS and CMS can make some measurements, and the ALICE detector (which we in ATLAS are living next door to) was optimised for precisely this purpose.

The concentrations of energy the LHC produces in proton–proton collisions are even higher. They take us back to energies above another threshold – the electroweak symmetry-breaking scale – above which the weak nuclear force is as strong as the electromagnetic force. This would be about 10^{-11} seconds after the start of the big bang. That is about ten zeros between the decimal point and the one. Before this, energies are even higher and frankly no one knows, though there are plenty of theories and a few constraints from data.

More about electroweak symmetry-breaking shortly, but back now to the quark–gluon plasma.

We managed to analyse and publish the data from the heavy-ion collisions remarkably quickly. On a Tuesday in November I was in the ATLAS control room doing my first shifts, monitoring and controlling some of the detector while we took more and more data from lead–lead collisions. During the quiet moments I was trying to write two talks, but more importantly I was trying to keep up with the collaboration review of a paper based on lead–lead data taken only days before.

These results were released and published a few days later.[73] They were a measurement of hadronic jets, again, but this time jets produced in heavy-ion collisions. And the interesting thing about them was actually the jets that weren't there.

Remember, hadronic jet is a spray of particles caused by a pair of quarks or gluons colliding, leading to quarks or gluons being knocked out of a proton, or in this case possibly a neutron, in the lead nucleus. Normally at least two jets would appear in such a collision, balancing each other so that momentum is conserved. A jet heads off in one direction, and an equal and opposite jet heads off in the other.

We had recorded some events where this happened, but we had recorded a lot of events where only one jet was visible. Momentum was still conserved, but rather than a second jet balancing the first one, a much more diffuse spray of lower-energy particles did the job instead.

What seemed to be happening is that, as intended, two lead nuclei were colliding and producing a brief soup of quarks and gluons. This is indeed the form of matter that filled the universe about a millionth of a second after the big bang. And it is within this soup that a single pair of quarks or gluons, one from each nucleus, were colliding with a very high energy.

In general, if such a collision happens near the edge of the soup, one quark only has to pass through a bit of soup to escape – that's the big jet. But in that case the other, going in the opposite direction, has to travel through lots of hot dense exotic matter. It gets scattered around and loses lots of energy into the medium it is passing through. That's the 'missing' second jet.

Hints of this behaviour had been seen before, at RHIC, but at ATLAS

---

[73] http://arxiv.org/abs/arXiv:1011.6182.

we had for the first time really measured it. And not only in a few events. We were actually getting a measure of how much 'soup' was being created in a given collision (more is made in collisions where the lead nuclei hit each other centrally, less if it is a glancing collision). We could see that in the events where more soup was created, more energy was being lost by the second jet, as one would expect. This meant we had a really good measurement of the nature of this soup. We were effectively using the quarks to study it, firing a quark through the material the early universe was made of.

Since those first results, ALICE, ATLAS and CMS have produced lots of results studying the effect in more detail, and also using other methods to study this weird new state of matter where quarks and gluons are no longer confined inside hadrons. The details of how much energy quarks lose as they travel, and how they lose it, have now started to tell us an enormous amount about the strong force, about this new form of matter, and about the early universe.

4.7 Putting the Higgs in its Place

On 31 January 2011, it was decided that rather than stop in the summer of 2012 as previously planned, the LHC would run all the way through to the end of the year. This made a lot sense. We hadn't yet published any results on searching for the Higgs, but we had learned a lot about the physics of 7 TeV proton collisions, the performance of our detectors, and the ability of the LHC itself to deliver data. Things were looking good, and from the projections we could now make, it seemed clear that if we extended the run until the end of 2012 we would have a very good chance of either finding the Higgs boson or proving that it didn't exist. So this is a good time to say more about why this would be such a big deal.

As already described symmetries are very important in physics, and the fundamental forces in the Standard Model are all gauge theories, generated by local symmetries.[74] The importance of the Higgs boson is connected to this, but in fact it is to do with breaking – or at least hiding – a symmetry rather than adding one. There are a few steps to go through to understand how this works.

Back in the mid-20th century, the first part of the Standard Model to be put in place was the relativistic quantum field theory of electromagnetic interactions, quantum electrodynamics (QED). Paul Dirac wrote down an equation that could describe electrons in a way that was consistent with special relativity, and eventually Richard Feynman, Julian Schwinger and Sin-Itiro Tomonaga proved that the theory was internally consistent. This internal consistency was in particular to do with a property called 'renormalisability'. The theory was plagued with infinities, coming from quantum corrections to the electron mass and charge.

The infinities arise, for example, from the fact that an electron travelling along can emit a photon, then reabsorb it. This makes a little closed loop. To calculate the probability of this happening in quantum field theory, you have to sum over all possible energies of that photon. The only condition is that the electron has to be the same after the loop as it was before. Unfortunately, because it is a closed loop, the energy that goes into the loop gets paid back, or cancelled out, at the end of the loop, so this condition isn't much of a constraint. In fact *any* amount of energy can flow around that loop! This is a disaster, because the energy in these loops gives a kind of infinite 'self energy' for the electron and ($E = mc^2$ again) this means the electron mass comes out as infinite. You do not have to be an experimental genius to notice that this is not the case for real electrons. In fact, we have measured the electron mass, and it is just over half an MeV.[75]

The obvious thing to do is replace the infinities by the measured value. Feynman, Schwinger and Tomonaga proved[76] that if you do this – replace

---

[74] See 3.5 Super Symmetry and Glossary: Gauge Theories (pp.96–9).

[75] 0.510998910 +/- 0.000000013 to be precise.

[76] Independently and in different ways, which Freeman Dyson later showed to be equivalent.

the infinite electron mass by the measured electron mass, and the infinite electron charge by the measured electron charge – then *all* the infinities in the theory disappear, and the theory (with those two experimental inputs) could make very precise predictions for all kinds of electromagnetic processes. This replacement of infinite mass and charge by finite, measured mass and charge is called 'renormalisation'. Feynman didn't like it much. He called it 'simply a way to sweep the difficulties of the divergences of electrodynamics under the rug'. But that is a direct quote from his Nobel lecture,[77] so it can't be all bad. This property of renormalisability is obviously quite important. Nobody wants infinities hanging around in predictions for real physical processes.

The same applies for the other fundamental forces. Several models for the strong and weak forces were proposed, but showing whether or not they were renormalisable was very difficult. Some of them clearly were not. Then Gerardus 't Hooft and Martinus Veltman showed that if a force is generated by a gauge symmetry, this is a necessary and sufficient condition for it to be renormalisable. This is important because it means that the trick of generating forces from symmetries is not just a neat way of getting a predictive theory, it is the *only* way. And it means that the sweeping of difficulties under the rug that Feynman grumbled about is not just a sneaky trick, it is built into the symmetries of the theory.

This was a big breakthrough and they duly won the Nobel Prize for it in 1999. But the Standard Model was not out of the woods yet. The problem is that the bosons generated by gauge symmetries were always massless. This is not a problem for QED, because the photon has no mass. It is also not a problem for the QCD, since gluons are massless. But it is a big problem for any theory of the weak force, since the W and Z bosons have a large mass. So, on the face of it, if you write down a gauge theory for the weak interaction, you have massless bosons, and if you add in the mass by hand to agree with observations, you break the gauge symmetry and end up with a non-renormalisable theory.

---

[77] He won the prize for this with Schwinger and Tomonaga in 1965.

Now, it is possible to hide symmetries. By this, I mean that you can have a completely symmetric theory that nevertheless gives rise to asymmetric situations. A favourite example is a marble[78] in a wine bottle.

If you look at a wine bottle from above, there is a circular symmetry around the centre of the bottle. That is, turning the bottle around an axis joining the centre of the neck to the centre of the base changes nothing.

Now imagine a marble in the wine bottle, but imagine the marble is bouncing around, perhaps because the wine bottle is being shaken (carefully). This is a representation of the early universe and the marble is a high-energy particle shortly after the big bang. Averaged over time the system is still symmetric because the marble could be anywhere in the bottle and is no more likely to be in one place than another.

Now place the wine bottle on a table and let the marble settle down. Because there is a bump in the base of the wine bottle, the marble cannot stop in the middle and has to roll off to one side. Once it has done this and come to rest, the situation is no longer symmetric.

This is an example of a hidden symmetry. All the physical principles governing the system – gravity, the kinetic energy of the marble and the shape of the wine bottle – are symmetric around the central axis. But the final configuration of the system once the kinetic energy has dissipated is not symmetric. The final configuration is the lowest-energy state of the system, what physicists call the vacuum or ground state, which just means the state with the minimum energy.

This trick can be deployed to evade the problem associated with having massive particles while preserving the necessary symmetries of the Standard Model. Keep the equations of system symmetric, but set them up so that they generate an asymmetric ground state – and that is the ground state in which the universe currently finds itself. In this ground state, fundamental particles, and especially the W and Z bosons, can have mass, but the weak interaction can still be a gauge theory.

---

[78] Actually, having broken my wine bottle when using it as an example at the Royal Institution, I recommend a plastic ball rather than a glass marble.

There is one more step to this, and since this is the crucial step made by the 'gang of six' (Brout and Englert; Higgs; and Guralnik, Hagen and Kibble) in 1964, we had better finish with it.

This idea of 'spontaneous' symmetry breaking had been tried before, and had worked in other contexts. But it came with a catch known as the 'Goldstone theorem', after Jeffrey Goldstone. This theorem states that when a symmetry like this is spontaneously broken, new massless scalar particles (i.e. particles with no spin) appear, as possible excitations of the quantum field.[79]

To see what this might mean in the wine-bottle analogy, imagine giving the marble, which has settled on one side of the bottle, a push, putting a tiny bit of energy into the system. If you push the marble towards the centre of the base of the bottle, up the hump in the middle, it will roll back. There is an 'excitation' of the marble possible here, where it rolls up and down the sides of the trough around the base. On the quantum scale, there's a minimum amount of energy you have to put in to make it do that, and for a quantum field, that minimum amount would be the mass. This oscillation corresponds to a massive particle.

But if you had pushed the marble along the trough so that it went around the base of the bottle, it would not roll back, it would keep right on rolling. In fact, precisely because of the hidden symmetry, all points around the base have the same energy, and you can make this marble roll around the bottom by putting an arbitrarily small amount of energy into it. The fact that there is no minimum amount of energy means this corresponds to a massless excitation – a massless particle. It is there because of the broken, or hidden, symmetry of the system, and Goldstone's theorem said this would always happen.

This is a big problem, because these massless scalar particles don't exist in nature. In fact, Yoichiro Nambu had already shown that spontaneously broken symmetry could allow hadrons (such as the proton and neutron) to have mass. But because of Goldstone's theorem, it also

---

[79] See Glossary: Fields, Quantum and Otherwise (pp.57–60).

predicted a massless scalar particle. Which just wasn't there.

What Brout, Englert, Higgs and the others did was show that when you have a gauge symmetry *and* a spontaneously broken symmetry present together, the massless scalar particle (from the broken symmetry) becomes incorporated into the gauge bosons (from the local symmetry). This simultaneously allows the gauge bosons to have mass, and eliminates the awkward massless scalar particle.

This was a big breakthrough, though it wasn't universally recognised as such at the time. This was probably at least in part because the idea found its place in the Standard Model in a way not really anticipated at the time.

When the 'gang of six' did this work, they were mainly thinking, like Nambu, of the masses of hadrons. This was back in the early 1960s, before the W and Z had been discovered and before the Standard Model was at all established. We now know that hadrons acquire their mass a different way. They are not fundamental, and their mass comes from the binding energy of their constituent quarks. Because of this, the Higgs boson is only responsible for about 1 per cent of the mass of everyday stuff. QCD is responsible for the rest.

But this 1 per cent is crucial, because it is the mass of the fundamental particles, and in particular, the mass of the W and the Z. Which, remember, seemed to be incompatible with them being gauge bosons.

In April 2011 I was in Edinburgh, home to Peter Higgs (since he left UCL in 1960), at the Science Festival. I love Edinburgh, right from the moment the train pulls in to Waverley Station. I love the fact that it has a daffodil-filled valley with a railway in the middle where most cities would have a river. I was visiting to take part in a discussion on 'Engineering the Large Hadron Collider', chaired by the comedian Robin Ince and with a talk from Lyn Evans, the project leader for LHC construction and the man who had counted us down to the great switch-on in 2008. My job was to be on at the beginning to try and explain *why* we had built it.

Hunting for the Higgs boson isn't really why we built the LHC. Actually, what we wanted to do was to understand why the electromagnetic force and the weak force have the same strength at high energies whilst being very different at everyday energies.[80] We knew that the difference in strength is caused by the fact that the W and Z boson have mass and the photon does not. Because they are massless, it is relatively easy to radiate and exchange photons, and they do not decay, so they can travel long distances (across the universe for billions of years, even). On the other hand, it takes a lot of energy to make a W or a Z, and even when you have one, it will very rapidly decay to other particles, meaning the weak force is short-range and, indeed, weak. But once the energy in an interaction is high enough that you can ignore the mass difference between the photon and the Z, the strength of the two forces is very similar. The energy scale at which they come together (if you are going up in energy) or diverge (if, like the universe, you are cooling down) is called the 'electroweak symmetry-breaking scale'.

This cooling down of the universe is like the marble settling in the bottom of the bottle. In the Standard Model, it is the quantum field equivalent of the bump in the bottom of the bottle that breaks the symmetry and introduces the masses of the W and Z, and in doing so introduces the difference in strength between the two forces. This is the Higgs field or, more correctly, the Brout–Englert–Higgs (BEH) field. It puts the universe into an asymmetric ground state, even though the underlying theory is symmetric.

Even if this model were wrong, the fact that the W and Z have mass would still be true, and with the LHC we would be able for the first time to do physics at energies well above the electroweak symmetry-breaking scale. Without the BEH field, the Standard Model would not work at these energies, so we would truly be in unknown territory and would hopefully get some clues as to the origin of mass, Higgs boson or not.

On the Sunday after the Edinburgh event with Evans and Ince, I chaired a talk with the *Guardian* science correspondent Ian Sample on his book *Massive*, about the history of hunting the Higgs. One audience

---

[80] Which does connect back to the Higgs in the end, bear with me.

question after Ian's talk was along the lines of 'If it is so important, ubiquitous and fundamental, why is the Higgs so hard to find?'

This is a really good question. The answer is that in a sense it's not the boson that matters. It is the BEH *field* that fills the universe, giving mass to the W, Z and other fundamental particles, that is the really important thing. If you think this BEH mechanism is correct, then every time you measure the mass of something you are seeing evidence for it.

On the other hand, this becomes simply a matter of interpretation, since the BEH theory has explained the mass, but has made no unique prediction for any new phenomena that you can test experimentally. Maybe some other theory could also explain the mass. In fact, this is pretty much why the draft of Peter Higgs' second paper on the matter was initially rejected by the journal *Physics Letters*.

He then went and added an equation that essentially says something along the lines of 'Well, if this field is there, you can also make waves in it and this will appear as a new scalar, i.e. spinless, particle.' These waves, or, again, quantum excitations, are not the massless scalar boson that Goldstone's theorem would have, but they are the leftovers from spontaneous symmetry breaking after the gauge bosons have had the rest. They still make a scalar boson, but one with mass. That is the famous Higgs boson, and that is why we have to see whether it is there or not. It was this prediction that made it possible to experimentally demonstrate whether the BEH mechanism was just a neat piece of mathematics or whether it really operates in nature.

On 22 April 2011, the LHC collided beams with a luminosity that passed the previous world record set by the Tevatron collider in 2010.[81] The intensity, or luminosity, is a measure of how many proton collisions

---

[81] The LHC reached 4.67×10^{32} cm^{-2}s^{-1}, compared to 4.024×10^{32} cm^{-2}s^{-1} from the Tevatron.

per second are happening in the LHC.[82] The units here are protons per unit area per second. So we were getting them at a faster rate than anyone ever had before.

If the Higgs boson existed, we were rapidly closing in on it. Data from the LHC collisions were streaming through our analysis software, and the tension was mounting day by day as we struggled to see what they might be telling us.

---

[82] See Glossary: Cross Sections and Luminosity (pp.72–4). This record of 4.67×10^{32} cm^{-2}s^{-1} corresponds to 4.67×10^{-7} inverse femtobarns per second, or (very roughly, depending on how long the machine could maintain this value and how quickly the team could refill the beams with protons after a run) 0.03 inverse femtobarns per day.

FIVE

Rumours and Limits

April–August 2011

5.1 Why Would a Bump Be a Boson?

By this time the LHC had been running at high energy for just over a year and we were deeply involved in analysing the data while greedily collecting more as fast as possible. It is about time I introduced some of the data distributions we were all obsessing over at this point. And indeed to unpack a bit what it means to measure something in a collider experiment like ATLAS.

The basis of measuring a distribution typically goes something like this . . .

First, you record as many interesting collision events as possible. A 'collision event', which we would usually just call 'an event', means in this context all the data we can get associated with a certain time-window in which a pair of proton bunches[83] passed through each other in the centre of the ATLAS detector. In LHC running to date, this has happened every 50 nanoseconds while the beams are on. So there are 20 million

---

[83] 'Bunch' is not a colloquialism, as in 'Hey what a crazy bunch of protons.' Bunch is technical jargon, as in 'The LHC is now operating with 1404 bunches.' The beams are not a continuous, uniform stream of protons, but a series of bunches, matched to the time structure of the radio-frequency accelerating fields of the LHC. Each bunch is a few centimetres long and contains 100 billion or so protons.

potential 'events' happening every second during LHC operation.

Most of these events are not 'interesting', that is either no protons actually collided with each other as the bunches passed through each other, or, much more likely, all the collisions were of a type that we have measured extensively already. We have sophisticated online selection algorithms (collectively called 'the trigger') to try and pick out the rare few interesting events. Some of these algorithms reside in hardware (application-specific integrated circuits – ASICs – and field programmable gate arrays – FPGAs – are the terms to bandy about here) and some in software running on a huge computing farm just next to the detector.

Data from a single bunch crossing arrive from the detector at a range of different times, not least because the detector is so large that a particle travelling at the speed of light takes up to 80 nanoseconds to travel from the collision point to the edge of ATLAS. So even before the particles from one bunch crossing have left the detector, the next lot are on their way. Then there is the time taken to read the information out of the detector and send it along cables to the computing farm. All the data have to be time-stamped[84] and all those time stamps have to be correct so that we end up building a picture of a single event: all the particles produced in a single bunch crossing. Of the 20 million possibilities every second, about 200 will be saved.

Next you have to 'reconstruct' the event. The events are stored at CERN, and also transmitted around the world to a network of computers (the LHC Computing Grid) to be reconstructed.

Reconstruction includes, for example, identifying the points where a particle passed through the silicon of a tracking detector, and performing pattern recognition to join the dots and make a particle track. These tracks are fitted to curves to extract the momentum that the particle had and extrapolated backwards to see where they started.

---

[84] Incidentally, the 'Timing Interface Module' that keeps the tracking and some other components of ATLAS in time with each other and in time with the LHC was one of the contributions of the UCL group.

Another part of the reconstruction code will identify pulses of energy in the calorimeter, where particles have stopped. These will be analysed to see whether they look like hadrons (which will have a long, deep shower of energy) or photons or electrons (where the shower will be more compact). What is the energy? And is there a track pointing to the energy? Photons should not have one since they are neutral, electrons should. And so on.

Some basic reconstruction will have been done already by the online selection algorithms, but here we do a more complete job, using the best calibrations and taking the time we need.

Next, you have to clean up the sample of events you have recorded and reconstructed. By this stage you really ought to have had a good think about what sort of events you want to measure for this particular analysis. With the trigger selection, anything not selected is lost for ever, so we err on the side of caution. By this stage, however, all the events are saved, so you can try different selections, optimise and re-optimise. Let's say that we are searching for the Higgs boson decaying to two photons. This is a rare decay for the Higgs (the Standard Model Higgs will do this much less than 1 per cent of the time – the exact number depends on the Higgs mass), but we can select for photons very efficiently, with a high degree of confidence, and measure them very accurately, which more than compensates for that.

So we want to identify photons. The results of the reconstruction will tell us whether there are any likely-looking candidates in our collisions. But the photons from Higgs decays would be higher energy than most – photons are the quanta of light, but the ones we measure here are about a billion times more energetic than the photons of visible light. They are also 'isolated', meaning that there aren't any other particles near to them in the detector. This is important, because photons can also be produced from the decays of a very commonly produced hadron, the neutral pion. These are just background noise if you are looking for the Higgs boson, but luckily they will most often occur in a hadronic jet, so surrounded by other hadrons, and they can be rejected.

One thing we have to worry about is that when the proton bunches collide, it is very likely (almost a certainty) that more than one pair of protons hit each other. The particles from all the pairs in a bunch that collided will be stored together, but we are only interested in the proton–proton collision that produced our photon candidates. We can generally remove the other stuff, either on a statistical basis (because we have measured the properties of the extra collisions, which will generally be minimum bias events, see section 2.2) or because we can see that some particles come from a different collision vertex from others. This is tricky for photons, however, since being neutral they leave no tracks and often can't be identified with a particular vertex.

Anyway, after selecting events with two isolated photons in them, and calibrating the measurement of the photon energy and direction, what next?

One can hypothesise that the photons were the exclusive product of the decay of a new particle of unknown mass. Under this assumption, we can calculate what the mass of that particle would be. The photons themselves each have zero mass, but their combined mass is not zero because they are moving away from each other very fast, so have a lot of energy whichever way you look at them. This is relativity in action. If they travelled in the same direction as each other, then you could in principle try to catch them up and their energy would get smaller the faster you moved in their direction.[85] But if they move away from each other, trying to catch one of them up increases the energy of the other one, and you can't get rid of the energy of the pair no matter how fast you go. And since energy is equal to mass times the speed of light squared, this means the pair has a mass too. Often this is called the 'invariant mass', because it is a quantity you can measure, or calculate, which does not vary depending upon how fast you are moving. The invariant mass of the photon pair does not depend at all on whether you try catching one of them up or just sit there in the control room drinking coffee. And, if they came from the

[85] This is the Doppler shift. The energy and wavelength of the photon change, although the speed of course remains the same – the speed of light.

decay of a new particle, this invariant mass of the photon pair will be the invariant mass of the new particle, since energy is conserved.

Then we can make a histogram of the data, that is a graph where we have the mass of the photon pair along a horizontal axis, and we divide it into 'bins', with each bin covering a small range of masses. Every time an event comes along, we put it into a bin along this axis according to the invariant mass of each photon pair we find. This way you get a graphical distribution displaying the number of events recorded at each mass value.

Even if there were to be a Higgs boson in there, most of the photon pairs will come from other sources. There are processes other than a Higgs that can produce two pairs of photons (for example, they could just be radiated off quarks). There will also be some fake photons in the sample. But these sources will generally have a smooth distribution of masses – there's no reason for them to cluster around a particular mass. Any photons from a Higgs boson decay, on the other hand, would be concentrated around a single mass, and so would show up as a bump in a smoothly falling distribution.

As time passed, we were collecting more and more photon pairs. There were fake, statistically insignificant 'bumps' in the distribution everywhere. This is where we just had to keep calm and carry on, despite the tension and the pressure for news – not just from particle physicists around the world but, bizarrely for a physics experiment, from the media.

While we collected more data, we were trying all kinds of tricks to make sure we were measuring the mass of the photon pairs correctly – calibrating and recalibrating the detector. If we saw a real bump, how sure would we be that we had the energy right? Or would we miss it, maybe wash it out by scattering the photon pairs around at wrong masses? White-knuckle time, really.

5.2 Crying Wolf

The Large Electron–Positron Collider (LEP) ran from 1989 to 2000 in the tunnel now occupied by the LHC. The LEP experiments made many of the measurements that established the Standard Model as a precise theory; amongst these were measurements that, when fed through the precise calculation framework of the Standard Model, indicated roughly what the Standard Model Higgs boson mass had to be (if it existed). Towards the end of the run, there was one of those heated decision points in particle physics: should the machine be turned off as planned, to start construction of the LHC, or should it run a bit longer at maximum energy to see if the Higgs boson popped up, right about the limit of sensitivity?

Extending the run would have cost a lot of money and would have delayed the LHC. But there were hints (how I hate hints) that the Higgs boson might be there, at a mass of about 115 GeV. The hints were just a few suggestive collision events, but a one-month extension was granted to see if more showed up. None did, so the machine was turned off. Several LEP physicists remained utterly convinced they had seen the Higgs and that its mass was 115 GeV. I have a colleague who bet several bottles of champagne on this, and another who gave a talk a year later in which he claimed they really had seen the Higgs already – and wasn't *entirely* joking.[86] Passions ran high.

Perhaps it shouldn't have been a surprise, then, that as the data were accumulating and we were all obsessing about the latest distributions (especially the two-photon mass I talked about in the previous section), a group of ATLAS physicists, amongst them some of the leading proponents of the 115 GeV 'LEP Higgs', got a rush of blood to the head – or some part of their anatomy – when for a while the data showed a small excess around the 115 GeV mass bin. An internal note making extravagant claims was distributed very widely within the ATLAS collaboration. It was clearly an overstated claim and in my opinion showed a lack of objectivity.

These are easy mistakes to make, but not so easy to email to thousands of colleagues. It should still have remained a slip-up between friends, part of the internal process by which ATLAS discusses such things and filters

[86] He later became a leading Higgs physicist at the LHC, so obviously – and very sensibly – recanted.

out those that aren't robust. Unfortunately, someone chose to breach collaboration confidentiality and post the title and abstract (and the internal author list, naming those would-be Nobel Prize winners who had beaten all their colleagues to the punch) on a blog . . .

The upshot was that many of us on the ATLAS experiment at CERN were a little busier than we anticipated over the Easter break.

The level of public and media interest in the Higgs search was high. The fear was that if too many false claims made the headlines, the interest would dissipate into cynicism, and if and when we actually had something to say, people would be too jaded to listen. On the other hand, to ignore the story would be to fail to engage, would risk us looking secretive, and would leave communication channels open to people who didn't really get the context. Having tried to ignore it as long as possible, when Channel 4 News called up I felt the story was already widespread enough that I had better help ATLAS respond.

I talked to Krishnan Guru-Murthy about it. I tried to explain that it really wasn't a hoax, but that the rumours were based on an analysis that had not passed the many levels of scientific scrutiny[87] required before anyone should get at all excited by it. It could fail at any stage, and given the manner of its release I expected it to do that, frankly. If it passed, it would be released by ATLAS and submitted to a journal.

At the time, all we really knew was that the results were overstated. The boson could have been at 115 GeV, but even if it was, the data at the time showed no significant sign of it. Apart from the embarrassment to the authors and the hopefully troubled conscience of the leaker, I think the whole episode was probably useful in educating the media and public about the difference between 'official results' and 'rumours'. Either can be right or wrong, of course, but stamping something as an 'official result' doesn't mean it has had policy approval, or commercial approval, or whatever else it might mean in the context of a political party or a company. What it means is that the collaboration of many physicists have

[87] See 3.3 Copenhagen.

done their absolute best to make sure it is correct, and correctly stated, and are all prepared to stand behind it. This is a real test for a complex result; it's the application of scientific method. It's not infallible, but it's a hell of a lot more reliable than a rumour . . .

I think we also gained some further confidence in dealing with the media on this. Scientists are sometimes too harsh in their judgement of journalists. The thing is, CERN is an exciting place now, as it was then. New data are coming in all the time. There are lots of levels of collaboration and competition. Retaining a detached scientific approach is sometimes difficult. And if we ourselves can't always keep clear heads, it's not surprising that people outside get excited too. I suspect we should be more forgiving of some of the excitable headlines – without, of course, encouraging them when they are misleading.

Reviewing is an occupational hazard of life in a big collaboration. The path from the first idea through to final publication of a result is arduous and full of potholes, trapdoors and mind-sapping disputes about commas and hyphens, often conducted at volume by a group of tired, angry people, none of whom have English as their first language. Even so, and the comma-and-hyphen element notwithstanding, reviewing is vital.

Meanwhile, in May there was another 'Boost' meeting, this time in Princeton. All good, but the best thing was that we had lots of data by then. We'd made the first measurements of some of the new jet substructure variables. Adam Davison and Lily Asquith, a postdoc at Argonne National Lab, showed the first ATLAS results. Miguel Villaplana, a PhD student at Valencia, actually showed pictures of the first highly-boosted heavy-particle candidates – two top quarks.

Glossary: Sigmas, Probabilities and Confidence

The business of trying to decide objectively whether or not to take seriously a bump in your data requires statistical analysis and agreed criteria. In particle physics we tend to talk about 'sigmas' as a way of quantifying this.

In this context, sigma (σ) is a parameter determining the width of a Gaussian, or normal, distribution. This is a bell-shaped curve that appears an awful lot in experiments because measurements often contain several independent possible sources of error. If you repeat a measurement many times, then the distribution of the results, with those independent and random errors in, will follow a Gaussian distribution,[88] with a particular width set by sigma, also called the 'standard deviation'.

Now, take a theoretical model – of, say, the two-photon mass distribution discussed in the previous sections – and a measurement scattered around the model, the actual data. If there is a visible bump in the distribution, we can use the 'sigma' idea to quantify how significant the bump is. In the Gaussian curve, made of many measurements, 68 per cent of the measurements are within 1 sigma of the central value, 95 per cent of them are within 2 sigma, and 99.7 per cent are within 3 sigma. So if we have estimated the sigma – the width of our error distribution – and we see a data point that is 3 sigma away from the background model, then on the face of it, if the data point really is just a background fluctuation, it must be one of the 0.3 per cent not lying within 3 sigma of the mean. Put it another way, there is only a three in a thousand chance that the background model alone describes the data.

This might sound like enough to get excited, and indeed conventionally '3 sigma' is deemed a sufficiently significant effect to be officially called evidence. But you have to be cautious. If you make a thousand measurements, the chances are three of them will show a 3 sigma effect, even if there is nothing but background noise in your data. This so-called 'look elsewhere' effect can also be estimated and factored into your confidence level. More crudely, the convention is

---

[88] This is the 'central limit theorem'.

that 5 sigma is what you really need to have to declare a discovery. This would correspond to one chance in nearly two million that your 'discovery' is just a background fluctuation.

In some areas of science it is more common just to quote the probabilities, or 'p-values'. We do that, too, often. But particle physicists do seem attached to their sigmas.

5.3 The Tevatron Goes Bump

The LHC was not the only game in town when it came to looking for bumps in distributions. The CDF experiment at the Tevatron proton–antiproton collider had a bump, which they actually reviewed and released[89] and which was published in a journal. This caused a stir and a special seminar. This is a more common situation in scientific research than unauthorised leaking. The result was (a) really important if confirmed, and (b) in need of confirmation before everyone could be certain of it.

What they had done was measure events in which a W boson was produced – subsequently decaying to an electron or muon, and a neutrino – and two hadronic jets were also produced. They then calculated the invariant mass of the pair of jets. At masses of about 150 GeV, there were more events than expected in the Standard Model.

A bump like this could be a sign that a new particle (a bit like the W boson but nearly twice as massive) was being made and decaying to two jets. There is no such particle known to science. It would be wonderful. It didn't fit very neatly into any expected theory, but within a day there were already several new theoretical papers eager to explain it. It would represent a massive breakthrough in our understanding of fundamental physics, though it would take quite some time to actually understand it!

There were plenty of reasons for caution, too. Firstly, there was the chance that it was just a statistical fluke. The paper quoted a one-in-ten-

---

[89] http://arxiv.org/abs/1104.0699.

thousand (0.0001) chance that the bump was a fluke. That's pretty small. However, lots of distributions like this get plotted. If you plot 1000 different distributions, the chances become about one in ten (0.1) that you'll see something this odd in one of them. Still, that's pretty small. And it's not clear to me that there really are 1000 different distributions as interesting as this one.

Another worry is that, quite apart from the statistical uncertainties, neither the data nor the predictions were exact. The energies of the jets were only known to within 3 per cent on average, for example. This is a systematic uncertainty. Systematic uncertainties are harder to evaluate probabilistically than statistical ones. If the energy scale of the jets is known to within 3 per cent, this means there is some reasonable probability that all the jet energies are wrong by 3 per cent.

If this was a statistical uncertainty, the 3 per cent would probably be the '1 sigma' uncertainty, meaning 68.2 per cent of the measured values would be expected to lie within 3 per cent of the correct value.[90] For a systematic uncertainty this won't necessarily be the case, though. They could all be uniformly too high by 3 per cent, for example. It is also hard to guess how probable that is. As with statistical uncertainties, if the '3 per cent' number is '1 sigma', there is a 68.2 per cent chance that the result is within 3 per cent of the right answer. If the errors are distributed normally, the 2 sigma point, which is the 95 per cent band, would be at twice 1 sigma, i.e. 6 per cent. But with a systematic uncertainty, the errors are unlikely to be normally distributed and it is very hard to assign probabilities to outcomes. CDF considered a lot of factors like this, and estimated that the probability of the bump being a false alarm is raised by a factor of eight (still only 0.0008, of course).

The other Tevatron experiment, D0 (D-Zero), followed this up later, and saw nothing, as did ATLAS and CMS. CDF also eventually performed a new analysis with more data, and also saw nothing. So like those Higgs rumours, it looks like this was a mistake of some kind. The Higgs rumour

---

[90] See Glossary: Sigmas, Probabilities and Confidence (pp.139–41).

hadn't been reviewed by the experimenters concerned (me and my colleagues on ATLAS), and failed miserably when it was. The CDF result was reviewed by the collaboration, and by a journal, and was a much higher-quality analysis. But even so, it was not reproducible and seems already to have found its place in the 'false alarm' folder.

It's worth going over some of these false starts, though. I think it gives a feel for how tense we all were, and of the pressures. It also gives flavour and weight to the kinds of cross-checks and studies that were going on within ATLAS and CMS all the time to try to make sure that if and when we did announce any progress on the Higgs search, it would not be retracted or contradicted later. But perhaps most important is the wider point about openness and reproducibility.

A second or third independent person looking over your results can see things you can't. Errors can be of omission, wishful thinking or just rubbish coding. In particle physics there is little commercial involvement (though careers are certainly at stake in these things). But when it comes to other areas of science, such as studying the effectiveness of a new treatment or a drug on people, it can be a matter of life and death. It is frankly astonishing that many health-care decisions (for instance whether to license a new drug) are based on data that are never released. Conclusions, even those based on good, honestly analysed data, can be wrong and can't be trusted unreservedly. They need to be reviewed by independent scientists and, ideally, reproduced. This is not even because they might be lying (though when big money is at stake it would be naive not to consider that possibility), but because if they haven't exposed their analysis to external review, they might be honestly wrong and we would never find out. Similar concerns or worse occur when libel laws are misused to silence criticism of analyses and conclusions. Either way, money is wasted and people suffer.

Reviewers are also good protectors of a night's sleep. One of the first papers I wrote was a measurement of the rate of production of jets of hadrons in photon–proton collisions, using data collected with the ZEUS detector. I made the plots in the paper myself, with my own code, and I became very aware of the enormous number of steps between the raw data and my result. I was also very, very aware of the fact that people on another experiment (H1) would likely repeat the result, and that physicists coming after me would measure it more precisely once we got more data. So any mistakes I made would eventually be exposed. But we were the first. No one knew the answer until we measured it.

In the early stages of the analysis, I would regularly find mistakes that would change the result by large factors. Things eventually settled down. The room for error got smaller. It got to the stage where I was tracking down tiny inconsistencies in the error bars, very minor things that, while some of them were still genuine mistakes, had essentially no impact on the conclusions of the analysis.

Even so, in the last week or two before publication, I woke up several times in the middle of the night in a panic, worrying about something I hadn't checked. The only thing that kept me (relatively) calm was the knowledge that the final plots had been made independently by someone else as well as me. In ZEUS we had a rule that required at least two independent analyses of any result for publication. In fact the 'second analyser' and I found several bugs in each other's analysis by comparing things at different stages. The chances of us both having made an identical massive blunder that would significantly change the final answer were pretty small. So I would go back to sleep – but still check when I got up.

Anyway, we were right and I am still very proud of that paper,[91] which was a significant step in understanding photons, protons and the strong interaction.

At the core of this behaviour is whether we want to know the answer, or whether we want to impose our view on the universe. In any science, and even more so in engineering, you'd better get the right answer. Nature won't pay any attention to a mistake and people may die. Finding an unwelcome right answer may appear to be a setback, but proceeding on the basis of a welcome but wrong result can be fatal.

---

[91] http://arxiv.org/abs/hep-ex/9502008.

I think one of the fault lines between science and politics is that this dynamic is less clear in politics. No matter their prejudices, politicians cannot change the science behind the consequences of carbon emissions, rape or vaccination. But when it comes to economic and social consequences, maybe they can. Persuade people to save, or spend, more money, you change the economy. Persuade people to support, or denigrate, universal health care, you change society. You can see why politicians, lobbyists or indeed drug-company executives may sometimes get confused about where the lines are.

Even where justified, using persuasion to change reality meets with varying success. But it has more chance of working in politics than it does in physics. Whatever worries you might have about quantum mechanics and the effect of the observer, don't try persuading gravity to give you a day off.

Glossary: Feynman Diagrams

A Feynman diagram is a particle physicist's best friend and worst enemy. As I have been describing physics processes in this book, most of the time I have had pictures of Feynman diagrams in my mind. Pretty soon (the next section, in fact) I am going to have to start actually drawing some diagrams and discussing them. So it is time to describe why they are so wonderful – and also why they should be treated with caution.

Feynman diagrams are cartoon representations of the actual equation you would use to calculate a quantum mechanical amplitude[92] for a particle, or several particles, starting in one state and finishing up in another. The initial state could be a pair of quarks inside protons in the LHC, the final state could be practically anything. As a simple demonstration, I'll use an initial state of an electron and a positron

[92] See 3.3 Copenhagen.

colliding in LEP, the accelerator that was in the same tunnel as the LHC but a decade earlier. And I will use as a final state a muon and an antimuon.

Here come the electron and the positron (time is running from left to right):

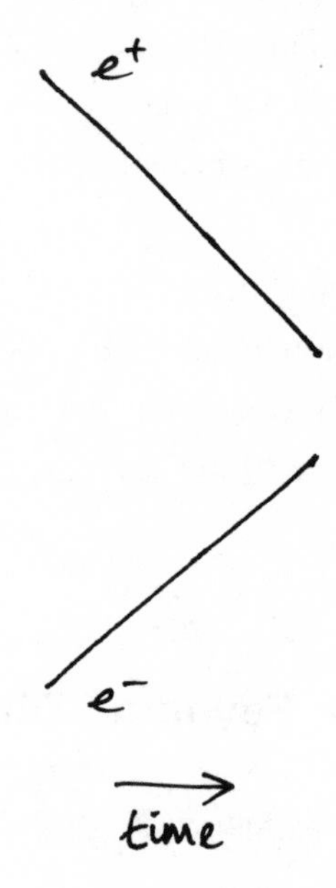

For each incoming particle there is a line in the Feynman diagram, and (though I won't write them down here) there is a corresponding term in the equation for calculating this. Now we let the electron and positron collide, and annihilate to give a photon:

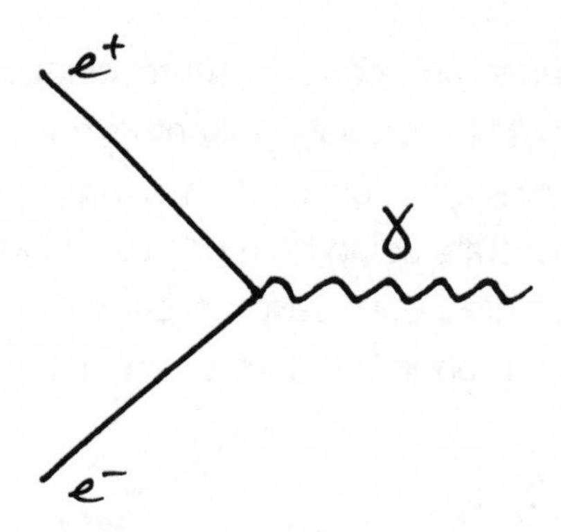

This vertex, where three particles meet, also has a corresponding term in the equation. In this case it is a very simple one – it is a number, the electron charge. The vertex is an electromagnetic interaction, and the probability of such an interaction occurring depends on the charge.

The photon continues for a while, and then decays, via another vertex, this time depending on the charge of the muon (which happens to be the same as the charge of the electron):

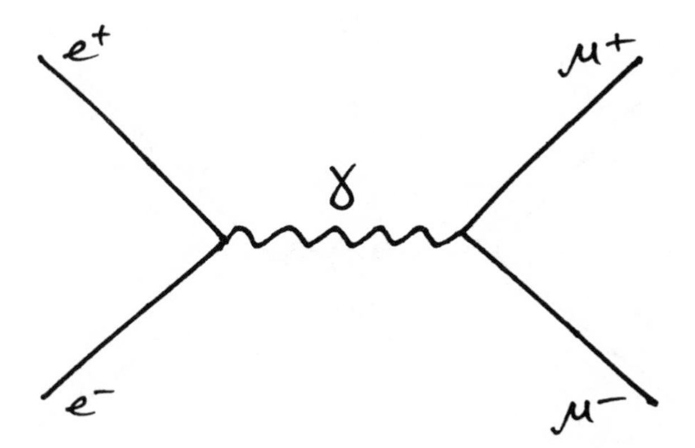

There is a term in the equation for each of the final-state particles (the muon and the antimuon), and for the wiggly photon line. This internal photon line is known as a 'propagator', and it is what we call a 'virtual' particle since while it is an important factor in calculating the amplitude, it can never be observed directly in the final state, so in a sense it is not real. The fact that it is virtual means that it can behave somewhat oddly. For instance, virtual particles do not have to have the same mass as their real-particle equivalents. This photon does not have to have mass zero – in fact in LEP it cannot have mass zero, because it has to be at rest, and it has to carry a lot of energy, so good old $E = mc^2$ means it has a mass that is bigger than zero.

To calculate the probability of this process happening one simply squares amplitude and combines this with a term describing how many electrons and positrons are incoming (the flux) and how many different final configurations are possible (the phase space). Not too tricky really, and the Feynman diagram gives a really intuitive, snooker-ball-style representation of the process.

And STOP, because that's where the caution is needed.

Feynman diagrams are amplitudes, and before squaring them to get the probability, you have to add up all the possible amplitudes. And there is another possibility we have neglected. As well as a virtual photon in the middle there, we might also have drawn a Z boson, like this:

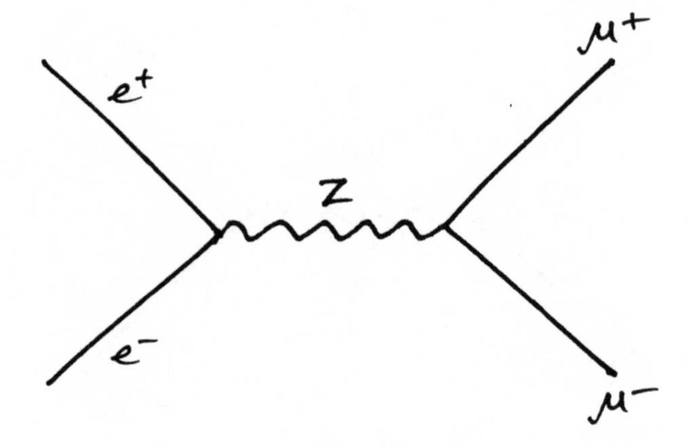

The initial and final states are the same, so the amplitude for this possibility has to be added to the previous one involving the photon to get the total amplitude. In fact, at LEP, because the collision energy was designed to be the mass of the Z boson – and therefore the Z propagator can have the correct mass – the Z amplitude was much bigger than the photon one.

To get the final answer we square the sum of these two amplitudes. The ordering (add first, then square, rather than the other way round) is crucial, because this is where the wave-like aspect of quantum mechanics is built in. Call the first amplitude, with the photon, P, and the second with the Z boson, Z. Add them to get the total T, so $P + Z = T$. The probability, T^2, is then equal to $(P + Z)^2$. This is different from the result I would get if I went in the wrong order, that is, first square the two amplitudes separately and then add: $T^2 = P^2 + Z^2$. You can see the difference matters immediately when you realise that amplitudes do not have to be positive numbers. Imagine the case where $P = 2$ and $Z = -2$. In the correct calculation, I get $T^2 = (2-2)^2 = 0$. This is an example of destructive interference between two amplitudes. If I do it wrong, I get $T^2 = 2^2 + 2^2 = 8$.

If you think of Feynman diagrams as real snooker-ball-like pictures of what is happening, you would never get interference effects, you would never get zero for the answer. You are neglecting the quantum nature of the interaction, and you will not describe the data.

A final twist on these wonderful diagrams . . . There are other amplitudes that one should add as well. For instance, there is no reason why two photons couldn't be exchanged, like this:

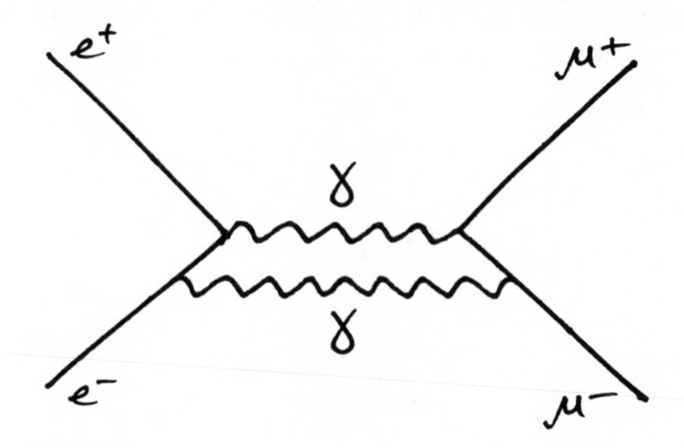

So we should add this amplitude too. And in fact an infinitely large stack of increasingly complicated diagrams is possible and should in principle be included. The fact that saves us from this, and means we just have to calculate a few diagrams, is that the number that comes into the equations with every new vertex – where three or four particle lines meet – is small. So every time a diagram gets more complicated, it also gets less important. This means that by calculating the first few diagrams, you can get a very precise answer, and the more complicated diagrams would just be a small correction to this.[93]

---

[93] This is perturbation theory, see 6.2 Perturbation Theory: Are We Covering Up New Physics? And it doesn't always work.

There you have them. Feynman diagrams: a beautiful and essential tool, very intuitive and helpful, but potentially deceptive if not treated carefully. Even the best physicists can be misled by Feynman diagrams occasionally.

5.4 W and Double W

In the summer of 2011, the data from the LHC started to show an effect that might, or might not, be the first indication of the presence of a Higgs boson. Most of this effect was due to the number of pairs of W bosons that were being produced in the proton–proton collisions. These extra W boson pairs could have been the result of a Higgs boson entering into the production process, like this:

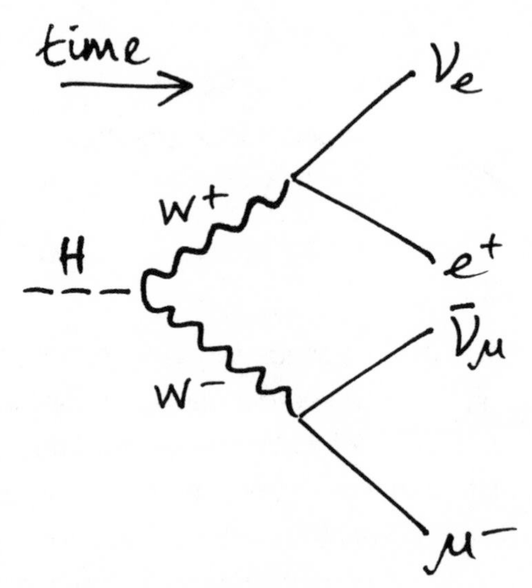

Diagram showing a Higgs boson decaying to a W+ and a W-, which then decay to leptons. (See Glossary: Feynman Diagrams, pp.145–50.)

It was tantalising to think this might be happening, but of course lots of caveats applied, both statistical and systematic.

If the Higgs boson were to have enough mass (that is, since $E = mc^2$, enough energy at rest), it would decay very quickly into a pair of W bosons. So if the Higgs was there, extra pairs of W bosons were one way we might first see it.

In contrast to the Higgs decay to two photons, however, a decay to W bosons will not lead to a bump. Generally, if you can measure the energy and momentum of the decay products you can reconstruct the mass of the original (in this case the Higgs boson). Unfortunately, the W bosons in question decay to an electron, or muon, plus a neutrino. The electrons and muons can be measured well (in the inner tracking detectors, the calorimeter and the muon system), but the neutrinos cannot. Neutrinos do not interact much at all with anything, including our detectors. To see any neutrinos, you need a detector much bigger and denser than ATLAS, and billions of neutrinos. So we don't see them. The missing momentum they carry away gives a clue, but it doesn't allow us to reconstruct a Higgs mass bump.

This is one reason the excess we were seeing was so susceptible to theoretical uncertainty. You don't need much of a theory to tell you whether or not there's a bump in your distribution. If the distribution has a significant bump in it, there's probably something there, whatever the theorists say. But we can't reconstruct a bump, so we just have to count WW pairs and see whether there are more than we expect. The theory tells us what to expect, but comes with errors.

About midnight on 14 June, the ATLAS and CMS experiments passed a major milestone six months early. The target at the start of the year, set for themselves by the people running the accelerator, was to deliver one inverse femtobarn of integrated luminosity to the experiments in 2011. The integrated luminosity determines how many proton–proton collisions the experiments can record. In line with the presentational sensibilities of the average particle physicist, this led to a Clip Art bottle of champagne featuring in the status reports the day after.

The round number was a largely arbitrary milestone. However, lots of the studies and projections done over the previous year used an inverse femtobarn as their 'baseline projection', so everyone knew very well that a lot of physics could be done with that amount of data, and that the accelerator was continuing to deliver at a great rate.

By coincidence, the first ATLAS paper on the search for the Higgs[94] was submitted on the same day; there was no sign of the Higgs, but it was easy to see that we would be closing in soon, and that the WW decay mode would be playing an important role.

5.5 Meanwhile in the Neutrino Sector

Even in 2011, particle physics was not all about hadron colliders.

Two major particle-physics labs are a couple of hundred kilometres south of Sendai in Japan. The J-PARC facility is right on the coast, in Tokai, and an accelerator there sends a beam of neutrinos to the Super-Kamiokande experiment, 295km away in Kamioka. This is the T2K long-baseline neutrino experiment.

The neutrino beam is measured in J-PARC, very close to the target in which it is produced. (You produce neutrinos by hammering protons into a lump of stuff. There's a bit more to it than that, but those are the basics.) Then the beam travels 295km, mostly unimpeded by the rock it travels through, to the Super-Kamiokande experiment. This is an enormous underground bubble filled with 50,000 tons of water and surrounded by 13,000 photon detectors. Some of the neutrinos interact here.

In the Standard Model, neutrinos are produced at the same time as an antilepton. This can be either an antineutrino, or a charged antilepton – a positron, an antimuon or an antitau. When they are produced along with a charged antilepton, they are labelled accordingly as an electron-, muon- or tau-neutrino. This label is called 'flavour'. Likewise, if and when they interact with the Super-Kamiokande detector, they mostly produce an electron-, muon- or tau-lepton, so you can label them by this at the end of their journey too.

---

[94] http://arxiv.org/abs/1106.2748.

However, neutrinos also have mass. And as they travel along, the mass is actually what is important, since it fixes the relationship between their energy and momentum. If you think of them as waves, which is the right thing to do in this case, the mass fixes their wavelength.

The odd thing is that the mass labels (call them masses m1, m2 and m3) do not match up with the flavour labels (electron, muon and tau). Because each lepton label is made of a mixture of mass labels, the neutrinos oscillate between lepton labels as they travel, because the different wavelengths go in and out of synchronisation as they travel. So the labels you see in Super-Kamiokande are not necessarily the same as the labels you see at J-PARC.

This phenomenon of neutrino oscillation has been measured before; it is the reason we know that neutrinos do have non-zero mass. In the old Standard Model, as it existed up until 1998, they had zero mass. There was an outstanding problem in physics called the 'solar neutrino problem', dating from the 1960s and stemming from pioneering measurements by the Homestake experiment led by Ray Davis and John Bahcall. This detector, in a mine in South Dakota, USA, measured the rate of neutrinos coming from the Sun and compared it to predictions based on what we know of the nuclear processes taking place there. The result was that there were not enough neutrinos. Various explanations for this were floated, including the possibility that the experiment was wrong (so it was repeated with independent techniques several times) or that our understanding of the Sun was wrong. No one could find a problem with Bahcall's model, but there was one rather terrifying possibility. Neutrinos and photons are both produced inside the Sun, and the number produced depends upon the rate of the nuclear reactions in the heart of the Sun. Photons produced inside the Sun scatter and re-scatter from the solar plasma and take thousands of years to escape, whereas neutrinos, because they only interact weakly, escape immediately. So a possible explanation for the deficit of neutrinos was that the core of the Sun was running down, and the sunlight we could see was the result of nuclear processes thousands of years ago, which had now slowed or stopped. If so, the Sun would be unstable, and we would be living on borrowed time.

Something of a relief, then, that another explanation turned out to be correct instead. The Homestake experiment only measured electron-neutrinos. If neutrinos oscillated to muon- or tau-neutrinos on their journey between the Sun and the Earth, then some of those that started off as electron-neutrinos would arrive as muon- or tau-neutrinos, would not be detected, and would therefore explain the deficit. This oscillation can only happen if at least some of the neutrinos have mass. So Davis and Bahcall were right, solar physics was right, but the Standard Model was wrong.

The breakthrough came in 1998 when Super-Kamiokande measured neutrino oscillations in neutrinos produced by cosmic rays hitting the upper atmosphere. An even more direct confirmation came when the Sudbury Neutrino Observatory, a big tank of heavy water in Ontario, Canada, used a different technique to measure all the neutrinos from the Sun regardless of their flavour, showing that it agreed with the expectations of the solar model.

Since then, neutrino oscillations have been measured more precisely using beams produced at accelerators and reactors. The first accelerator experiment to do this was the Main Injector Neutrino Oscillation Search, or MINOS. (A good name that doubles as the Minnesota–Illinois Neutrino Oscillation Search, given that the beam is fired from Fermilab in Illinois to the Soudan mine in Minnesota.) T2K was part of the next generation of these experiments.

Back at ICHEP in Paris, Eric Zimmerman from the University of Colorado had shown some of the first neutrino events from the T2K experiment. These were being analysed as more data were collected, and the result had been eagerly anticipated. However, a catastrophic earthquake struck Sendai in March 2011, just before a planned seminar to announce the first T2K results on neutrino oscillations. In fact, a series of seminars around the world had been planned. The lab suffered serious damage and for a worrying period all the websites and email addresses at the big high-energy physics labs, KEK and J-PARC, went offline. Though everyone in the neutrino experiment hall of T2K was evacuated safely, the seminars were postponed until the results could be announced in Japan first. In June, finally, the results were made public. T2K had seen indications of muon-neutrinos turning into electron-neutrinos.

The Standard Model can be modified, and has been, to accommodate neutrino masses, but this is a major change. It's a case of 'The Standard Model is dead – long live the Standard Model.' But it is a profound change, and has taken a long time to absorb properly. I still find myself examining PhD theses where the introduction states that 'In the Standard Model, neutrinos are massless . . .'

In the post-1998 Standard Model, the way neutrinos oscillate can be defined using a matrix containing three angles. By 2011, two of these had already been measured by several experiments, including MINOS. But the third was unknown, and could have been zero, in fact. T2K was seeing indications (at the level of a couple of sigma) that it was not zero. That would be quite profound news, if confirmed.

Apart from the fact that neutrinos themselves are important – they are fundamental and they are everywhere, and their presence affects how the universe formed and developed – this question goes way beyond neutrinos alone. In nature, all the fundamental matter particles come in three copies, or generations. Like the neutrinos, quarks also have these three copies. The mass of each generation is heavier, so muons are basically heavier copies of electrons, and tau leptons are heavier still. Likewise, top quarks are heavier than charm quarks, which are heavier than up quarks.

Also we know (from measurements of the decays of the Z boson at LEP) that if there are any more generations out there, they are much, much heavier – even the neutrinos. Or else they are weirdly different in other ways. It looks like the fundamental particles of nature come in three and only three copies. The Standard Model does not predict this should be so, it just seems to be that way.

This is pretty odd, and is connected to one of the big open questions in physics – where did all the antimatter go? Since everyday matter, every atom, is basically made of the first generation alone (up and down quarks, and electrons), the other two copies, or generations, seem a bit superfluous. But there is a really intriguing hint in this of some deeper theory. Three copies are *just* enough to allow matter to be different from antimatter. The effect can occur when you mix up three or more copies, but mixing up two, or only having one, is no good. Experiments have shown that such mixing, and matter–antimatter differences, do occur between quarks. What was unclear was whether this might also happen with neutrinos. Going back to the mixing angles: if the third mixing angle were zero, then it couldn't happen, because we would only have two-way mixing. If, and only if, the third angle is non-zero, then matter–antimatter differences could happen amongst leptons as well as quarks. Given that we are made of matter, not antimatter, such differences are important. How the universe, at least as far as we can see it, got to be made of matter and not antimatter is one of the outstanding questions of physics and cosmology. The matter–antimatter differences we do see in the interactions of quarks seem to be too small to explain it all. So if the neutrinos also show some differences, that would not only be a big new piece of physics in its own right, but it might be a clue as to what is missing in our understanding of the whole relationship between matter and antimatter.

Anyhow, the first step is to see whether the neutrinos mix three ways or just in pairs, because that's a necessary condition for the kind of matter–antimatter asymmetry we would expect in the Standard Model. At this point, T2K was saying most likely they are, though the uncertainties were still too large to be sure. Meanwhile, physicists in China and Korea were working on it with nuclear reactors . . . of which, more later.

5.6 Quantum Fields and Missing Quotes

I was asked to give a CERN seminar on 7 June about ATLAS measurements of jets. I always learn something when I have to give a seminar, even if the audience don't. One of the things I learned this time is that not everything is on the Internet yet. Of course I knew this really, but I needed reminding.

Jet measurements tell us a lot about the strong nuclear force, QCD, which holds the atomic nucleus together. We talk a lot about how the LHC may find 'new physics', by which we often mean new particles or new forces, or even wilder stuff like extra dimensions of space–time. However, we also have a lot of new physics to learn about the forces we already know of, especially the strong force.

Strong forces in general are tricky. They can lead to consequences that are impossible (or at least very difficult) to predict with current techniques. Things like the masses of hadrons, and the way quarks and gluons are distributed inside them, can be understood to a very large extent using different approaches to calculating with QCD, but all of them need some experimental inputs, even though in principle it should be possible to derive them from the theory alone. This derivation is tricky exactly because the interaction is strong. The emergent properties of strongly interacting quantum field theories are complex, fascinating and worthy of study.

Anyway, I remembered a quote I had used in a talk nearly ten years before from Martinus Veltman, who won the Nobel Prize in 1999 with Gerardus 't Hooft for work on fundamental forces. The quote was something like:

> If the Large Hadron Collider finds a Higgs boson and supersymmetry, nature will have missed a golden opportunity to force us to understand strongly interacting quantum field theories.
>
> *Nobel Prize winner Martinus Veltman (possibly)*

What he meant was, if there is a Higgs boson and supersymmetry within reach of the LHC, our current techniques for solving quantum field theories will be 'good enough' to understand them. But if there were no Higgs, a very likely outcome is that even the weak and electromagnetic forces become strong, and therefore badly understood, at LHC energies. In this scenario, if we want to understand anything, improving our understanding of strong forces in general would be essential.

Of course, my point in the talk was that we should do this anyway, since we know one of the fundamental forces is strong.

I wanted to use the quote again, so I went searching for the source on the Internet, like you do. I knew the words might not be exact, so I tried the usual tricks of different combinations in Google and so on. No joy.

In desperation, I went searching for my own talk, in which I knew I had used the quote. I had to dig out an old password to get into the site, but in the end I tracked down my slides. Only to find in there the approximate quote, with the words:

> (Paraphrased . . . couldn't find the exact quote this morning.)
>
> *Jon Butterworth (being sloppy ten years ago)*

Damn.

I know I met Veltman at Nikhef, the Dutch national subatomic physics lab in Amsterdam, a few months before that talk. He was associated with the lab so they were celebrating his Nobel, and by coincidence I was giving a seminar there, and Susanna had come along for a weekend break. We got invited to the party. We drank quite a bit of champagne and, a bit star-struck and urged on by Susanna, I got the great man to sign my *Rough Guide to Amsterdam*. I suppose it is remotely possible that I actually got the quote direct from him at this party. If so, I hope the paraphrase is accurate. Either way, the sentiment is still correct, and I used it again in the CERN seminar.

5.7 Exclusion Limits: The Boson Fights Back

As the conference season rolled on in the summer of 2011, the Large Hadron Collider continued to deliver shedloads of data (or inverse barnloads, perhaps). We continued telling people that unauthorised leaks should be ignored because the results had not been reviewed. But this of course just meant that the pressure mounted to get the real, official results reviewed and approved as soon as possible. An organised chaos of analysis, cross-checks, comments and approvals continued around the clock (literally, since ATLAS is a collaboration crossing many time zones). This was not just about the Higgs results, but also searches for other stuff that might happen, and measurements of what was actually happening in our collisions. The last of these took up most of my time, since that was the job of the group I was coordinating.

Absorbed in those measurements as I was, I couldn't ignore what was happening in the Higgs searches. In fact, some of the measurements my group was making (for example the measurements of WW production) were essential cross-checks and inputs for the search. Eventually we had our results ready for the next big conference – in this case the European Physics Society (EPS) meeting in Grenoble.

As is usual for scientists, I worry a lot about bias. The prospect of going to all the effort of building an experiment to find an answer, but getting the wrong answer because of subjective bias, is a nightmare. This is true of any experiment, but it's very stark for one the size of the LHC and for a question as important as whether there is a Higgs boson or not.

Once I knew the ATLAS results, but before they were public, I wrote an article about them. Of course, I didn't publish it until after the results were released, but at the point of release we would also have the independent results from CMS, and I wanted to put down my thoughts about our data without using the CMS data as a cheat sheet that might bias me. Here's what I said:

> It is Wednesday evening, after a day of sitting in discussions where the last batch of ATLAS results were approved to be shown at EPS. These included our Higgs searches. I can only reveal the results on Friday, which is why you aren't reading this before then. But I am writing it beforehand for a good reason, which I'll explain at the end.
>
> . . . With the amount of data we have analysed, we should have ruled out possible Higgs bosons with masses between about 130 GeV and 200 GeV, and also between about 320 GeV and 460 GeV. If we had done this, it would be a big extension of the previous

exclusion limits and would really be squeezing the possible hiding places for the Higgs . . . All this, of course, is if there is no Higgs boson.

. . . What the data actually say [is] we have excluded at 95% confidence a Standard Model Higgs with a mass between 155 GeV and 190 GeV, and similarly between 295 GeV and 450 GeV. Some of this range was already excluded by the Tevatron experiments, but this is still a big advance. In fact around 290 GeV we are doing 'better' than expected, which could of course just be to do with fluctuations in the data.

But it gets even more interesting about 155 GeV. Below here, we are not doing as well as we should.

This means one of three things:

(1) We were unlucky, and an upward random fluctuation in the background messed up our sensitivity.

(2) We did something wrong and failed to find it yet (these are still preliminary results, though they have been reviewed by the ATLAS collaboration).

(3) A Higgs boson, or something very much like it, is lurking in there somewhere below 155 GeV and is starting to emerge, blinking, into the light.

I am writing this now because I want to say what I think about our own data before I see what the Tevatron experiments, or our friends across the LHC, CMS, are going to say. By the time you read this, I will probably know. But here you get my judgement of what our stuff means, independent of any bias from other experiments.

If [at EPS the other experiments] show exclusion limits going down below 155 GeV, this means ATLAS was unlucky (or wrong) and the Higgs has less room to breathe.

If they show exclusion limits which they didn't expect to go below 155 GeV, this means they are not sensitive enough to say anything about the ATLAS result in this region.

If they show exclusion limits which, like us, they *expected* to go below 155 GeV but which *don't*, then the chances that this is the first sign of the Higgs boson emerging into the light are increased.

It's not decisive, but from the ATLAS data alone the odds have shifted in favour of the Higgs being around. When CMS and the Tevatron show their data, the odds will either shift back again, or shift further in favour of the Higgs. I don't know yet. But in contrast to some previous occasions, it's now getting really interesting.

The ATLAS Higgs results are being reported by Kyle Cranmer (New York University) at 15:00 CEST (after which I will post this article), just before CMS. All the Higgs results from the Large Hadron Collider will be summarised in the plenary later by Bill Murray (STFC Rutherford Appleton Laboratory). They are the result of a massive amount of work by hundreds of people on the LHC machine and on ATLAS.

I am putting this in here because I think it gives a sense of the real uncertainty at the time – the caution, but also the excitement.

As I've mentioned, and whatever the lunatic black-hole-armageddon fringe might say, when it comes to particle-physics results lives don't generally hang in the balance. But many of the same issues – statistical confidence, systematic bias, blinding of experiments – apply to medical trials, or climate research, where lives are at stake. In all cases there are usually plenty of massive egos and vested interests involved. Real knowledge about how things work emerges from rumour, claim, counterclaim and honest doubt.

Regarding those results I wrote about before the EPS – what actually happened? Which of those three options turned up? Well, the Tevatron experiments were not sensitive to the standard-model Higgs in the relevant mass range. Their sensitivity stopped at 148 GeV, at the time. However, CMS were indeed sensitive, and saw something similar to ATLAS (without having seen our results first, just like we had not seen theirs). So that improved the odds further.

Next we needed to reduce the uncertainties.

Firstly, the statistical ones. The excess could have been just a random upward fluctuation. Imagine tossing a coin to see if it was fair. If the first four throws all came up heads you might become suspicious, since you would have expected two heads and two tails. But this is not a very significant result. The chances of getting four heads even with a fair coin are one in two to the power of four, or one in 2 x 2 x 2 x 2 = 16. And the chances of getting four the same (either four heads or four tails) are twice this (one in eight). So you would have to carry on tossing the coin a lot longer before you were more convinced (to, say, one in a thousand, or 99.9 per cent confidence) that it was biased. This is an example of statistical uncertainty on the confidence with which you can say something – in this case, 'the coin is biased'. The uncertainty shrinks as you take more data – toss the coin longer. So obviously we would continue to collect data from more proton–proton collisions at the LHC and to analyse them – the equivalent of tossing coins – so that we could see whether the Higgs was in there, biasing the result.

The second type of uncertainty is systematic. These uncertainties can come from how well (or how badly) we understand our detectors. For example, if an electron hits it, how often do we actually see it? And how well do we really measure its energy?

ATLAS and CMS are completely independent detectors. So not only do they double the statistics (shrinking the first kind of uncertainty), they also have independent, and very different, systematic uncertainties arising from how the different detector technologies are understood. So seeing things in both experiments really builds confidence. Also, the more data we get, the more control experiments we can do to test and improve our understanding of the detectors.

Unfortunately, not all systematic uncertainties are independent. Of those that are not, the most important is the uncertainty in the theoretical calculations. The theory tells us what a Higgs boson of a given mass should look like, and also what the background – non-Higgs production events – should look like. CMS and ATLAS both rely on this. When we say 'we have an excess of events' we mean we have more events than one would expect if there were no Higgs boson. And we are both using the same theory to do this. So if it is wrong, we could both see a false signal.

The theory we have is actually very good. However, it is only a theory. To reduce the systematic uncertainty here, the method is similar to the way we reduce detector systematic uncertainties, but also involves theorists. We have to do more control experiments, measuring the production of different kinds of particles at LHC energies, and see how well the theory describes them. And if it doesn't, we have to find out why and fix it. In this case especially we needed to measure the production of W bosons, since most of the excess we were seeing at that time was in events where two W bosons were produced.

On to the next round, and the next conference. It was August 2011 and every big conference was getting updates on progress in the big Higgs hunt. I had missed Grenoble, but was heading to Mumbai.

5.7 Mumbai

The tea on the plane was excellent.

I never thought I'd write that sentence, but I suppose if you are ever going to get excellent tea at 30,000 feet it will be on a BA flight to India. India were losing to England at cricket, but the taxi driver still talked about little else all the way through the suburbs. This was a proper car: the little three-wheeled autorickshaws teem through the outskirts but are banned from central Mumbai, and I suspect from the airport too.

I travel a lot as a particle physicist, but I'm not really very widely travelled. Most trips are back and forth to Geneva, but even before the Large Hadron Collider induced that semi-regular commute I mostly stayed in Europe, with the occasional trip to North America or Japan. Basically, the richer and more industrialised bits of the world, I guess. Perhaps that isn't so surprising. Even I would not put major particle-physics facilities at the top of the wish list in the first stages of developing an economy.

That said, particle physics would feature quite early on. The activity has lots of economic benefits and is an essential part of any national research infrastructure that attempts to address big questions about the nature of the physical world we inhabit. Plus, the subject is by necessity very collaborative. Rolf Heuer, Director General of CERN, said in his final address at the European Physical Society meeting that the 'E' in CERN now stands for 'everyone', not just for 'European'. If you want to do physics at the energy frontier, then at the moment you have to go to CERN, wherever you are from. And to carry on doing science at this frontier in the future will require global collaboration.

India has been part of this for a long time. The famous Indian physicist Satyendra Nath Bose gave his name to the generic class of particles with integer angular momentum – bosons – of which the Higgs was at this point a still-hypothetical example, but the W and Z bosons (and the photon) were well established. The Tata Institute in Mumbai has a long history of involvement in fundamental science. In 2011 it was hosting the Lepton Photon conference. Cue visa hassles, immunisations, tea and taxis, and the realisation that I had been jetting about inside a bit of a cocoon.

At Lepton Photon conferences there are no parallel sessions. The talks are all plenary 'rapporteur' talks – a great place to get an overview of the field. At Lepton Photon 1999 at Stanford, USA, I gave a talk on photon structure.[95] The thing I remember most from that meeting, though, is hearing Saul Perlmutter describe the results on supernova brightness that were part of the new and mounting evidence for dark energy in the cosmological standard model. Those results won the Nobel Prize for Perlmutter, Brian Schmidt and Adam Riess in 2011.

To be honest, I also remember the Napa Valley tour and the excellent wine selections we had for the lunches and dinners on the Stanford campus, for which I think Jo-Anne Hewitt[96] and the other organisers still deserve credit.

---

[95] It doesn't really have a structure, but in an interesting way.

[96] See 3.5 Super Symmetry.

The other Lepton Photon meeting I attended was in Uppsala, Sweden, in 2005. I gave the talk on experimental studies of QCD. Gavin Salam (as seen in *Colliding Particles*) was giving the theory talk on the same subject. I remember being asked many times whether I was related to Ian Butterworth, a former research director at CERN, and thought it odd that no one asked Gavin if he was related to Abdus Salam (Nobel Prize-winning theorist). As far as we know, neither of us is related to either of them, anyway.[97] I also remember very long queues in bars for the toilets because apparently (some) Swedes don't believe in gender-specific toilets, thus making men suffer along with women in an admirably egalitarian fashion, but without actually providing additional capacity. They should have put up notices: *Dear Men, Please – like women must do normally – plan your toilet visits 30 minutes in advance to avoid discomfort.* Also the fish was brilliant, especially for breakfast. A herring breakfast beats even English bacon and eggs, in my opinion.

Given the depth of my analysis of cultural differences – wine, toilets and herring – I suppose tea and cricket in India were to be expected.

There was a slight tone of disappointment at Lepton Photon in the reports of the Higgs updates, which was understandable. A month before at EPS, ATLAS and CMS both had hints in the data that, while not statistically very significant, could have risen in significance at this meeting. However, even though we added more data, the significance did not rise – in fact it dropped a bit. But again, not significantly. All consistent with statistical noise in either direction, up or down.

At EPS, it became a little more likely that the Higgs was around in the mass region 130 to 150 GeV. At Lepton Photon it became a little less likely. The 95 per cent exclusion went down to 145 GeV. At higher masses, huge swathes of possibilities had been excluded. But still we had said nothing about the region between 115 GeV and 125 GeV.

We would do so soon, however.

---

[97] Although it turns out that Ian Butterworth was my PhD supervisor's supervisor's supervisor. My great-grand-supervisor, I suppose.

5.8 Which Leads, Theory or Experiment?

One recurring theme when you talk about science in public is a perception that fundamental physics is too theory-led; that we are obsessed with proving beautiful, reductionist theories when it would be better if we just explored. And that we spend too much time arguing about untestable things. This is not a criticism to be dismissed lightly, and some of the time, for some physicists, it is almost certainly a fair one. However, even though 'hunting the Higgs' was explicitly a major part of the reason for building the LHC, I'd make three arguments against the idea that there is too much theory.

1) Thought experiments

Theoretical discussions about currently untestable things can flush out conflicts and inconsistencies in our understanding. Black holes, for example, in one sense amount to an 'extreme thought experiment', highlighting a conflict between three amazingly successful theories, or 'laws of physics', if you prefer. Quantum mechanics, gravitation and thermodynamics have their laws, and their underlying picture of the universe. They have credibility by virtue of each being able to describe a vast range of phenomena, from steam engines through planets to the central processing unit in your computer. In a black hole, these laws come into apparent conflict. The territories of three giants overlap. By thinking through the contradictions that arise, we can find gaps in the theories, develop new understanding, and in the end hopefully derive observable predictions that could be used to test such understanding. It is a hugely worthwhile exercise, unless you are utterly uninterested in understanding how things work or in benefiting from such understanding.

2) Electroweak symmetry-breaking

To some it appears that the Large Hadron Collider is a disproportionate investment of time, money and expertise in chasing some theorists' dreams. I disagree, of course. While the Higgs is the headline, the LHC is genuinely exploring new territory for whatever might be there. The energy frontier (or, if you like, the short-distance frontier – we study nature at smaller-distance scales than anywhere else) remains a frontier of knowledge, whether Peter Higgs says it is or not. Plus, we have very good reason, from experiment alone, to think that this part of the frontier is special. Look at this plot:

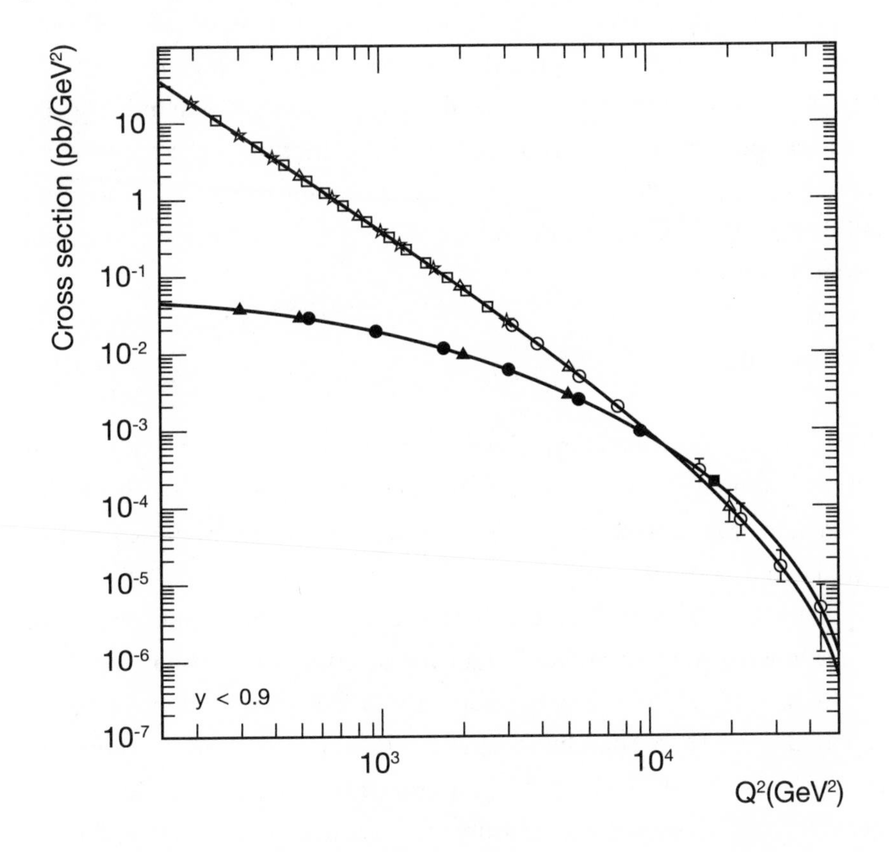

Data from the ZEUS and H1 experiments at DESY, Hamburg.

What this shows is essentially the probability of an electron bouncing off a proton, with the energy of the bounce increasing as you go from left to right. The clear points show the times when it bounces because of its electric charge – the electromagnetic force. The solid points are the times when it bounces by swapping a W boson – the weak force. You can see that at low energies (towards the left) the electromagnetic bounce is much more likely. But at high energies (on the right) the weak force is just as likely to be responsible as the electromagnetic. There is a symmetry between the two forces that is restored at this energy. These are data from the HERA experiments. Measured. No theory. (Actually the curves are theory, but you can ignore them if you wish.)

The LHC, for the first time in the history of science, was allowing us to explore properly above that energy, into the region where the symmetry holds. Our theory says the Higgs breaks the symmetry. But even without that theory, you might think exploring physics above this fundamentally important energy scale is an exciting thing to do and might tell us how these forces work and why they are sometimes the same and sometimes different.

3) *Gloating*

Finally, the LHC data had even by this stage, after just of a year of running, led to a bonfire of theories. While big ideas like supersymmetry or extra dimensions had not been completely disproved, many options for them have been closed off.

On the last day of the Lepton Photon conference, the Mumbai monsoon let rip in a big way. The noise of rain was so loud at lunch that you could hardly hear the SUSY theorists weeping into their curries after the results reported by LHCb that morning. LHCb is the one of the four big experiments on the LHC but has featured least in this story so far, largely because it is designed to measure rare particle decays (especially of hadrons containing b quarks, hence its name) and as such does not participate in the hunt for the Higgs. However, it does have very good sensitivity to physics beyond the Standard Model. The SUSY theorists were not really quite in tears, but the LHCb talk by Gerhard Raven had dealt their favourite theory a blow. He reported a beautiful measurement of a specific decay of the B_s meson (a hadron containing a beauty quark and a strange quark). This agreed very well with the Standard Model, which, despite the confidence of some theorists, does not include SUSY. Previous, less precise measurements had been a bit away from the standard model and the difference could have been due to SUSY. It wasn't. More hopes dashed. SUSY is even more slippery than the Higgs, though, and rumours of her demise are exaggerated.

Also at the meeting, a friend of mine, Jenny Thomas, reported on recent and future long-baseline neutrino experiments such as T2K. One thing she had to say was that yet another 'hint of new physics' – this time from her experiment, MINOS – had gone away. This was an apparent unexpected difference between the behaviour of neutrinos and antineutrinos, which had first been reported at ICHEP in Paris the previous year. The measurement was interesting but not very precise and, as with LHCb, more precision had again vindicated the Standard Model, to the disappointment of many.

Jenny made the point that we often start taking such anomalies too seriously too soon. People talk about 90 per cent exclusions, or 10 per cent probabilities that some piece of data is consistent with the Standard Model. But 10 per cent is not a small number. You will have many 10 per cent chances coming up if you do a lot of science. And as we saw in Mumbai, results are flooding in all the time, from the LHC but also from elsewhere. It's fine to be interested, excited even, but we have to keep a realistic attitude. For the Higgs search at least we talk about 95 per cent exclusions. But '3 sigma', which is the conventional standard for real evidence, is 99.7 per cent probability.[98] And 5 sigma, conventionally a discovery, is 99.99994 per cent. And there are good reasons for these conventions. Even with such high levels of certainty, odd things happen and mistakes still get made.

It is true that many of us have our favourite theories, but in the end the data decide, and as an experimentalist I seriously enjoy making myriad bright ideas face the music. We had waited a long time, theorising and guessing. Now, at last, we were able to look at some more of the answers.

---

[98] See Glossary: Sigmas, Probabilities and Confidence (pp.139–41).

SIX

First Higgs Hints and Some Crazy Neutrinos

September–December 2011

6.1 Faster-than-Light Neutrinos: A Case Study

We were now building up to the CERN Council meeting in December. We would be required to give an end-of-year report on the LHC results, especially (of course) the Higgs boson search.

The CERN Council is the governing body of CERN. CERN was set up by 12 European states[99] in 1954 to help rebuild European physics after the Second World War, when many of the world's best physicists had fled Europe, often initially for the UK, where they were in general immediately encouraged to head for the USA, which they did.[100] And many of them then helped build the bombs that were dropped on Hiroshima and Nagasaki.

CERN was set up to be explicitly non-military. Several countries not aligned in the cold war (including, for example, Switzerland) were founder

---

[99] Belgium, Denmark, France, the Federal Republic of Germany, Greece, Italy, the Netherlands, Norway, Sweden, Switzerland, the United Kingdom and Yugoslavia.
[100] Thankfully, of course, the UK would never be so stupid and unwelcoming to asylum seekers and talented would-be immigrants these days. Would it?

members, and even at the height of international tensions cooperation with the Soviet Union and other Warsaw Pact countries continued. The summer school I attended in 1991 was organised jointly by CERN and its nearest Soviet equivalent, the Joint Institute for Nuclear Research (JINR). I got to visit Dubna, the big Russian lab near Moscow, and we spent two weeks in Alushta on the Black Sea, in the Crimea. A few weeks later, Gorbachev was staying in a hotel quite near ours when the coup took place that triggered the break-up of the Soviet Union.

The Council is made up of two representatives from each CERN member state, as well as the Director General and his immediate team. It is also often attended by representatives from observer states and states applying for admission. The UK delegates are a senior civil servant from the ministry responsible (Business, Innovation and Skills, at this time) and the chief executive of the relevant research council (currently the STFC). When one of these can't make it, I often attend instead. As well as overseeing the running of the laboratory in Geneva, the CERN Council is also mandated with the 'organisation and sponsoring of international cooperation in the field'.

I find it easy to imagine Director General Rolf Heuer's face when, just as we were building up to the December reports that might be decisive in terms of the Higgs, another CERN-related experiment knocked on his door and asked to give a seminar claiming to have measured a particle travelling faster than the speed of light.

The particles were neutrinos, and the experiment was OPERA.[101] At the same time as the seminar, they submitted a paper to the arXiv[102] and a journal, so we could read and review the evidence behind this extraordinary claim.

OPERA measures neutrinos produced from the SPS. The SPS (Super Proton Synchrotron) is one of the accelerators in the CERN accelerator

---

[101] Oscillation Project with Emulsion-tRacking Apparatus. Dodgy acronym, dodgy experiment?

[102] arxiv.org/abs:1109.4897.

complex. In a previous incarnation as the Super Proton–Antiproton Synchrotron, it was responsible for the collisions in which the UA1 and UA2 experiments discovered the W and Z bosons in 1983. Carlo Rubbia and Simon van der Meer received a Nobel Prize for this a year later, and in retrospect, though CERN had made important breakthroughs before, this looks like the moment it became a really world-leading laboratory. The SPS is now also an injector for the LHC. So, a very useful piece of kit.

The neutrinos are regularly fired 732km under the Alps and beyond to the Gran Sasso National Laboratory in Italy, where some of them are detected by the OPERA experiment. This experiment is designed to measure how the neutrinos change their properties as they go. They are trying to measure the appearance of tau-neutrinos, in fact. But they also have some very precise Global Positioning System (GPS) position and timing measurements. So they know the distance the neutrinos travel, they know how long it takes them and they can therefore measure the speed. Since neutrinos have a tiny mass, the speed should be very close to the speed of light. But they measured neutrinos arriving early – thus travelling faster than the speed of light.

If this result were valid, it would be an amazing breakthrough. The speed of light, *c*, as a maximum speed limit is built very deeply into the maths of how we understand the universe.[103] It is one of the foundations of the theory of relativity, which precisely describes all kinds of physics, including the physics behind the GPS systems used to make the measurements and the accelerators used to make the beam. So it can't just be thrown out. Some better theory would have to be found that contained and extended Einstein's edifice.

It might be hard to imagine what such a theory might be, but pretty quickly the fans of 'extra dimensions' theories came up with some thoughts whereby the neutrinos take a short cut across another dimension. Sort of the equivalent of a two-dimensional Londoner living on the surface of the globe taking a short cut to Sydney via the Earth's core.

---

[103] See 2.2 Minimum Bias.

Fun to think about, but all a bit premature. It was just a single result, and the OPERA people themselves could hardly believe it. The paper looked like quite a careful publication, and the collaboration had reviewed it at some level, though some members did say publically that this had not been as thorough as they would have liked. Dario Autiero, the OPERA physics coordinator, gave a scientific seminar at CERN on 23 September, when they received and answered a lot of questions. This is the way science is done, and I didn't blame them for speaking to the press too, and being excited about the potential implications of their data. What should they have done, kept it secret? Imagine the conspiracy theories if they had done that. 'CERN suppresses evidence that Einstein was wrong!!' I already get plenty of green-ink letters and emails shouting about that. Preventing OPERA from showing the data they wanted to show would have justified some of those.

I also did not and would not blame the media for getting excited by this. What should they have done, ignored it? Gratifyingly, people are interested in physics and this was a proper story. It may have been overexposed and in some cases overhyped, but this was a genuine scientific debate, going on in real time about an intriguing result. It was not a manufactured controversy. As long as people appreciated that, maybe seeing science done in public could become a new, educational, spectator sport. Ideally even a participation sport. That would be a good thing, surely?

The big question of course was, 'Is it correct?' What might have gone wrong? It looked like a careful study, the result of years of work by a pretty big team, so to fire off 'It must be wrong' comments without due care seemed unfair, even though I must admit that was my immediate reaction. Jim Al-Khalili had no such qualms, promising to eat his boxer shorts if it was right. But in the spirit of doing science in public, I tried to explain my main concerns with the results.

This was to do with the time distribution of the protons at the source in CERN. The protons hit a target and ended up producing neutrinos. OPERA also measured the time distribution of neutrinos arriving at Gran

Sasso. They fitted one distribution to the other, and when they lined up, that gave them the time of flight, and thus the speed. Most of the power of the measurement came from the leading and trailing edges of the pulse – that is, the arrival times of the first and late neutrinos in the pulse.

The claim was that fitting to the distribution gave an accuracy of about 10 nanoseconds (6.9ns statistical and 7.4ns systematic uncertainty). This seemed bold to me. But the main worry I had was they seemed to assume that the shape of the time distribution should match exactly. There was no allowance in their estimate of their systematic uncertainties for the possibility that they might not, and it wasn't hard to imagine reasons that the shape, as a function of time, of the neutrino pulse at Gran Sasso might not be a true reflection of the shape of the neutrino 'turn on' and 'turn off' at CERN. To me that looked like an odd, possibly serious, omission. For example, at CERN, where the neutrinos were produced and measured initially, all the protons were included in the time profile. By the time it got to Gran Sasso the beam fanned out to cover an area much bigger than the OPERA detector, so OPERA would only see neutrinos from part of the beam. So any correlation between the production time of the neutrinos and the angle they were produced at (which would determine whether they actually got to OPERA or not) could distort the shape, leading to an uncertainty in the fit and hence an uncertainty in the speed.

I ended up sketching this on a napkin for a BBC documentary presented by Marcus du Sautoy that was made impressively quickly after the first results came out. I thought it was an interesting discussion to have, though I didn't claim it to be a debunking of the result.

I'm glad I didn't. A few weeks later, the OPERA collaboration submitted an updated version of their paper. The most important change since their first version was that a new test had been done. Now, instead of firing long blobs of neutrinos, CERN had been firing short pulses, just three billionths of a second long. So now to measure the time, the OPERA physicists did not need to know the shape of the pulse, they just needed to know which pulse the neutrino came from. They redid the study, and after seeing just 20 neutrinos they came up with the same answer: the neutrinos still appeared to be travelling a bit faster than the speed of light.

That was one test survived, but there were other tests that needed doing. To be honest, this was such a remarkable result, with such profound implications if correct, that we needed at least one completely independent experiment to check it. The other long-baseline experiments – MINOS in the USA and T2K in Japan – started gearing up to do this.

The scientific discussion continued in public. According to their inclinations, scientists were assuming it must be wrong, thinking about what could be wrong, thinking about how to test it, or wondering what might be the implications (for physics and Jim's shorts) if it were right.

Now, unsurprisingly, it was wrong. More surprisingly, and embarrassingly, it was wrong in a way that no one outside the collaboration could have worked out from the paper. There was a cable connector, converting between electrical and optical transmission, that had not been plugged in properly. Once that was fixed, the answer came out to be consistent with the speed of light. This was a disappointing end to a fun story. It should have been possible for the OPERA physicists to work out that there was a possible single point of failure like that. It should have been checked before they went public. The leaders of the experiment who made those decisions stepped down.

There are three reasons why I think it is still worthwhile going through all this here, even if it is a bit of shaggy-dog story.

One is that it is worth thinking about the opposite situation. If the measurement had agreed with expectations, and showed neutrinos travelling at or below light speed, how carefully would we have examined the uncertainties? Would that connector have been checked? How much time would I have spent reading the paper? And if the result had been expected but wrong, would we ever have found out? This would be an example of confirmation bias in science – where if something aligns with expectation, you count it and credit it and stop looking for mistakes. I have measured many things that 'agree' with the Standard Model. I have also made measurements that didn't agree with the Standard Model, but this always made me suspicious, and I went looking for errors. I have always found them – either in my own measurement, or in the way we were using the theory in the comparison. We check all our measurements carefully, but I cannot honestly say that we check them equally carefully whether they agree or not with the Standard Model. If enough people repeat honest experiments, this will still be corrected in the end, but it takes time and is a subtle biasing effect.

A second reason is just to point out, again if necessary, that scientific progress is not as neat and tidy as it might appear with hindsight. I am writing an account of some scientific history here. Without giving away the ending, I must say there is a chance it might end up looking like an inevitable progression towards a triumphant conclusion. This is not how it appeared at the time. Science has many blind alleys and mistakes. In the end they are irrelevant and so are forgotten, if and when the data are eventually unambiguous. But we kid ourselves, and do little credit to the process or the history, if we only remember the correct experiments.

There is a third reason.

There are quite a few people out there who are obsessed with 'turning physics on its head' (often professional physicists and journalists), or at least with proving Einstein wrong (amateurs, usually with green pens), or who are just generally outraged at science's claims to some kind of objective truth and special status (the more mediocre philosophers or sociologists, mostly, but sometimes people with a political or religious agenda at odds with the reality emerging from data). The idea that new data can disrupt science is a wonderful thing, and the fact that it can still happen even with all the confirmation biases I've already discussed is precious. In fact, there are huge incentives for scientists to produce disruptive, paradigm-shifting data. You will be famous – briefly, like OPERA, if you are wrong, but for a long time if you are right. But this does not mean that when we find something new, we throw out what we already know.

There were plenty of overviews in the media of what we hoped to learn from the LHC, many giving lists of theories we might find evidence for, or disprove. The sort of theories listed – supersymmetry, extra dimensions, whatever – typically addressed some of the problems or omissions in the Standard Model. They generally postulated new phenomena – particles or forces, on the whole – that might be visible at the LHC. A list of new theories provides short cuts for experimentalists like me. If you know what you are looking for because some theory tells you, you can do a focused and efficient search. But if you don't find whatever it was, the usefulness of the search depends on how seriously the theory was taken in the first place. This is a somewhat subjective criterion. Producing results that exclude some forms of supersymmetry, for example, has been of great interest, since many people take supersymmetry very seriously as a candidate for an extension of the Standard Model. As some kind of measure of this, if you write such a paper, it will get cited lots of times.

This approach on its own is somewhat dissatisfying to me as an experimentalist, and is not the whole story. I don't see my career as consisting of chasing evidence for particular new theories, or trying to confirm the Standard Model, for that matter. The most important thing is to make measurements of what actually does happen in the new energy regime to which the LHC gives us access, and to use these measurements to challenge and improve our understanding of nature. This involves comparisons to theoretical predictions, of course, but the approach is different. The measurements are usually quite independent of the theory, in that they stand whether or not a particular theory turns out to be the right one.

Many LHC papers, perhaps too many, consist of unsuccessful searches for evidence for specific theories of physics beyond the Standard Model. Many others, however, consist of measurements more independent of the theory. The two approaches are complementary, and there is also a grey area where searches are made for rather generic, and so less theory-dependent, new phenomena.

If evidence for one of these new theories turns up, or we find something that doesn't fit the predictions of the Standard Model, one could say the

Standard Model is wrong. It surprises people sometimes to know that such an occurrence would be greeted with pretty much universal joy by particle physicists.

The scientist and author Isaac Asimov wrote a nice little essay (a letter, really) describing how the framework within which we understand nature is refined by science. Theories are thrown out, new ones replace them, but the process is not circular. Theories are not just intellectual fashions. In science, each successful new theory corresponds to a more complete set of natural phenomena, thus giving greater understanding, being more useful, and in this sense being more true. Asimov gives a fascinating discussion of the merits of different theories about the shape of the Earth: flat? spherical? oblate spheroid? pear-shaped? At one point he says:

> When people thought the earth was flat, they were wrong. When people thought the earth was spherical, they were wrong. But if you think that thinking the earth is spherical is just as wrong as thinking the earth is flat, then your view is wronger than both of them put together.

On the scale of a few miles, the Earth is quite flat, if you average over mountains and valleys. That's quite a sophisticated observation. But it is slightly wrong – there is a small curvature, which is of course crucial. Similarly, if the Higgs boson didn't show up soon, that would mean the Standard Model was wrong. Even if the Higgs boson showed up, the theory could be (and probably would be) 'wrong' somewhere else eventually. But over the distances and energies we'd studied so far, it could only be slightly wrong, because it described the existing data so well.

Just as with the small curvature of the Earth, any small deviations from the Standard Model we see could be critical for our understanding. They would lead to the replacement of the Standard Model by some bigger, better theory.

That would not mean the Standard Model was a waste of time. It is much more right than what we had before. And the new theory would be even more right. As Asimov also said, it probably makes more sense to describe previously successful but now discarded theories as 'incomplete' rather than 'wrong'. The new theory would have to do everything the Standard Model did in terms of describing existing observations, and also describe the new observations. It would be more complete, and in this specialised sense, more true. Hence the joy.

In just the same way, anyone 'proving Einstein wrong' really has to incorporate the results of Einstein's theories where they agree with data, show where they are incomplete, and make them more complete. This is not a matter of fashion. I am old enough to remember Manchester, my home town, when purple flares were ubiquitous, then despised, then everywhere again, and then once more not. This is not science. Science advances, and is different from fashion, or philosophy. It amazes me how many academics study the sciences as a phenomenon without being able to acknowledge or understand this, the most fascinating and distinguishing aspect of this human activity.

I had been booked to discuss the neutrino controversy at the Cheltenham Science Festival as part of a panel discussion chaired by Jim Al-Khalili, secure now in his boxer shorts. By the time the festival came along, the controversy was over and the results unambiguously wrong. Yet the discussion went ahead in front of a packed hall and covered a lot of this ground. Physics can be interesting even when it's wrong, I guess.

6.2 Perturbation Theory: Are We Covering Up New Physics?

The big question we are addressing at the LHC could be summed up as: Does the Standard Model of particle physics work at LHC energies or not? And 'LHC energies' are a significant step forward, since they are above this energy of electroweak symmetry-breaking, where two forces unify and where the mass of the W and Z bosons, and possibly all the other fundamental particles, originates.

If the Standard Model works in the new regime, there would be a Higgs boson but not much else new. If it doesn't, there might not be a Higgs but there must be something weird and new going on. There is a key question lurking behind this, which is: How well do we really understand the predictions of the Standard Model at these energies? This isn't an easy one. In general we can't solve the Standard Model exactly. We use approximations. Most of these rely on the fact that the 'coupling', that is the strength of the fundamental forces,[104] is not very large.

The strength of a force can be expressed as a number. If it was 0.1, say, then the chances of two particles interacting would be proportional to 0.1 x 0.1 = 0.01. But if a third particle were involved it would be 0.1 x 0.1 x 0.1 = 0.001, a fourth would be 0.0001 and so on. This means when the coupling is small, you can ignore the contributions that involve more than say four particles – they are just a small perturbation on the main result, because they are multiplied by 0.1 x 0.1 x 0.1 x 0.1 x 0.1 = 0.00001. They don't change the result much. This is an example of 'perturbation theory', widely used in solving many problems in physics and chemistry. It is accurate if the coupling is small – that is, if the force is weak.

This approximation is not always valid. The bits when it does not work mostly involve the strong force, QCD. That's why it's called the strong force. (We don't intentionally obfuscate, it's tough enough as it is.)

For example, some aspects of how quarks and gluons are distributed inside the protons we collide can't be calculated from first principles.[105] Neither can the way the quarks and gluons turn into new hadrons in the end. We have some constraints from our theory, we have basic stuff like the conservation of energy and momentum, and we have a lot of data from other places. But we can't use perturbation theory. The coupling number gets near to one, and 1 x 1 x 1 x . . . = 1. This means no matter

---

[104] The number associated with each vertex in the relevant Feynman diagram – see Glossary: Feynman Diagrams (pp.145–50).

[105] See 4.5 Inside a Proton.

how many particles you include in your calculation, you don't converge on a reliable answer. In the end we have to make educated guesses, or models. And these are always adjustable.

A serious worry, then, is that we might adjust these models in such a way that we actually cover up exciting new physics. To avoid this happening, you need to have calculations of what you know, made with perturbation theory, linked up to models of what you don't know very well, which you have some freedom to adjust. I think of this rather gruesomely as a skeleton of hard predictions inside a squidgy body of best guesses. The body can change shape. You can push in its stomach or squeeze its cheeks relatively painlessly, but it has two of each kind of limb, and you really know about it if you break a bone.

Anyway, marrying the squidgy models to the rigid perturbation theory is mostly done using computer programs known as Monte Carlo event generators. These not only encode much of what we know about what happens when particles collide, but they are also an invaluable tool in designing new experiments and working out how your existing experiment is responding to data. 'Monte Carlo' is an allusion to the fact that, like roulette, they use a lot of random numbers.

There's an interesting bit of sociology of science around all this. As a theorist you can sometimes lose out from being involved in one of these generators. You can have a paper with thousands of citations and people will say 'It's only software' or 'It's just a Monte Carlo thing,' whereas with a similar number of citations in string theory you might stride the world like a colossus, despite the fact that the generator will describe data, whereas string theory struggles to predict anything remotely measurable.

Monte Carlos are not the only way, but in general they are part of an effort to understand the implications of the Standard Model and to try and get it to make as many precise predictions as possible. Given the relative size of the communities, this is an effort on a scale comparable to the effort of building the LHC itself. As some recognition of this, in 2011 the American Physical Society's J.J. Sakurai Prize was awarded to three theorists, Bryan Webber, Guido Altarelli and Torbjörn Sjöstrand, who work in this area. The citation read:

> For key ideas leading to the detailed confirmation of the Standard Model of particle physics, enabling high energy experiments to extract precise information about Quantum Chromodynamics, electroweak interactions and possible new physics.

This made me very happy because, for one thing, two of them are close colleagues, and for another, calculations and code written by all three of them are essential to understanding pretty much everything we are doing at the LHC, including making sure we don't cover up any new physics by mistake. As we continued trying to quantify and reduce our uncertainties in the Higgs search, looking for the crucial 3 sigma evidence or 5 sigma discovery, comparisons between Monte Carlo generators and the incoming data were carrying on around the clock.

6.3 Counting Sigmas

And so, on 13 December 2011, we made our end-of-year report to the CERN Council. And, it turned out, to a sizeable fraction of the world's media. By the time I thought to get in the queue, the main auditorium at CERN was completely full. People had been eating their breakfast in there in advance of the afternoon talks. So I went into the overspill room, the 'Filtration Plant', which was also set up for the media. All the buildings at CERN have numbers, but some lucky ones have names, too. The Filtration Plant is actually (now) quite a nice meeting room, whereas the Pump Room is a large, cold garage with a few fold-up chairs and a projector in the corner. You have to know these things when you run a working group at CERN. The meeting-room booking system defaults to Pump Room.

The Filtration Plant was full too, with about half-and-half journalists and physicists like me who had not got up early enough to find a space in the main auditorium. The talks, from our glorious leader, Fabiola Gianotti, and Guido Tonelli, the head of CMS, were to be webcast, and there was a feed into our room. The atmosphere was intense. Not only were journalists prowling for comments and sound bites, but there was a real anticipation amongst the scientists about the results.

I knew what Fabiola was going to show, of course. We had a hint of a signal, mostly in the two-photon mass spectrum,[106] but it was less than 3 sigma (the conventional point at which one declares 'evidence for . . .') and a long way from the conventional 5 sigma threshold for a declaring a discovery. This in itself was exciting, of course, and I was beginning to think there might actually be a Higgs boson after all. But what I was really keen to see was what CMS had. Rumours had inevitably been spreading, and we expected that they had some hints too, similarly suggestive but inconclusive. There was a chance, though, when one saw the details, that the hints would either be contradictory (which would be disappointing, even a bit worrying) or would line up so well that they might at least push the combined result close to or over the 3 sigma level. So today might conceivably see the first official declaration of 'evidence for a Higgs boson'.

What no one was prepared for was the great font controversy. There were gasps when Fabiola began her talk – she was using the much loathed Comic Sans. Somehow people could grasp this more easily than the content of the slides, and Twitter went into some kind of paroxysm. Much more importantly, the CMS hint neither entirely strengthened nor contradicted the ATLAS result. This was a bit frustrating, both for the physicists and for the journalists trying to work out whether there was a story there and if so, what it was.

Well, there was a story, anyway. Long after the rest of the room had cleared, I was standing in the Filtration Plant being interviewed over a link by Jon Snow on Channel 4 News while Mike Paterson meta-filmed the interview for the *Colliding Particles* films. Jon Snow seemed very interested in the lack of linearity in our progress. I'm not sure we were

---

[106] See 5.1 Why Would a Bump Be a Boson?

being non-linear here, just experiencing an erratic pace of progress along a line. But of course there was no guarantee it was the right line, and hopefully the previous rumours, CDF bumps and neutrino mistakes have given you some idea of why we remained cautious. Cagey, even.

Channel 4 News was not even the end of the day. With a few friends I went back to the UCL flat in Meyrin to celebrate, but also to have a teleconference with Brian Cox, who was on stage at the Hammersmith Apollo with Robin Ince doing their *Uncaged Monkeys* show. The audio was not good, and I think mostly people just got the impression of happy physicists and drinks, but the audience reacted enthusiastically despite the hiccups (technical and beverage-induced). The evening after it went a bit better, and John Womersley, the CEO of STFC, joined us and was a little more coherent. After all the technical troubles and battles with funding that Brian, John and I had been very much involved with, it was rather hard to believe that Brian was standing on the stage talking about gauge theory to about 4000 people and video conferencing to CERN to talk to miscellaneous scientists, including the head of our research council, and everyone was cheering. And we hadn't even discovered anything yet. Some kind of dream, but a lovely one.

That day we also had the ATLAS Christmas dinner. It was in the Pump Hall, which is OK for events like that. I was too exhausted to write anything sensible for the *Guardian* on this at the time, so I wrote a limerick that summed up the main conclusion:

A physicist saw an enigma
And called to his mum 'Flying pig, ma!'
She said 'Flying pigs?
Next thing you'll see the Higgs!'
He said, 'Nah, not until it's five sigma!'

Seriously, I had been a confirmed Higgs sceptic until now. This was the point at which I started to think it was probably there. Of course, that's when you have to be most careful and watch your judgement.

6.4 Higgs Boson in Coupling Shocks, Bumps and Stupidity

It might sound as though the situation was not too different from that of the previous summer. What made this result – inconclusive though it was – more compelling was the fact that it was in the two-photon distribution, where a bump is expected, whereas in the summer the 'hint' had been in the WW decay mode of the Higgs. As I have already described,[107] this mode is not very good for telling us the mass of any candidate Higgs boson there might be. The neutrinos, from the decaying W bosons, carry away too much information and we don't see them. The WW decay mode still featured in both ATLAS and CMS, contributing to the hints. But the main interest was now focused on two other ways the Higgs can decay and therefore show up in our detectors.

Both of these decay modes can tell us the mass of the Higgs, if it is there. Both should show bumps and so are less susceptible to theoretical and systematic uncertainties. This is why I was now taking the statistical evidence (which said it was likely but not certain a Higgs existed) seriously. Statistical uncertainties are much easier to assess than systematic ones.

Both the decay modes we were focused on were a bit weird, though, for different reasons.

The first decay mode is the Higgs decay to two photons, which I have already talked about.[108] The weirdness here is that the Higgs boson is famous for – or was deduced from, or invented to explain, according to taste – mass. Fundamental particles get mass by interacting with the BEH field, of which the Higgs boson is an excitation. By the same token, then, the Higgs will generally decay to heavy things. The more massive they are, the more likely it is that the Higgs boson will decay to them, because it interacts most strongly with them. Conversely, things with no mass don't interact with the Higgs.

---

[107] See 4.3 Prospecting and Surveying.

[108] See 5.1 Why Would a Bump Be a Boson?

So why photons? Photons are quanta of light. They have no mass. They should not interact with the Higgs boson!

Indeed, the Higgs decays to photons very rarely. If the Higgs boson mass is about 125 GeV, and you make 1000 of them, fewer than ten will decay to two photons. Most will decay to bottom quarks, in hadronic jets (jets containing a bottom quark are called b-jets). But these are very hard to distinguish from other collision debris that doesn't involve a Higgs. Jets, even b-jets, are cheap at the LHC. Picking out the Higgs decays to bottom quarks is the subject of the paper I had written with Gavin, Adam and Mathieu just before the LHC started,[109] and as I write, we still haven't done it successfully. It will have to wait for the higher-energy data.

In contrast to b-jets, pairs of high-energy photons, not surrounded by other stuff, are much rarer and can be measured more accurately. But since the photon's mass is zero, the Higgs really ought not to decay to photons at all. And it does not, directly. It has to go through a loop of some other particle, as in this cartoon:

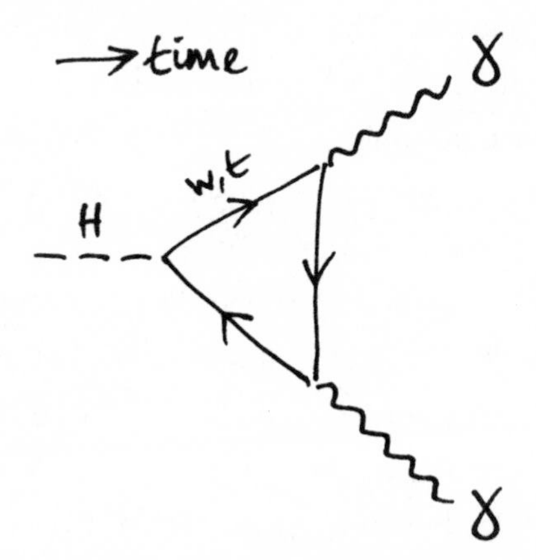

Feynman diagram of Higgs decay to two photons

This is why the decay rate is low. Referring back to the perturbation theory idea, every vertex where three particles meet carries with it a number less than one – the coupling – and the loop has more of these

---

[109] See 1.7 Boost One.

(three) than direct decay would (one). So probability of a decay to photons is small, but it can still happen. This is fine. In quantum mechanics, anything that can happen has to be included in your calculation. There might even be other new particles we've never seen going round that triangle, though in the Standard Model it's usually a W boson or a top quark.

The other important decay mode at this stage is the decay to pairs of Z bosons. If the Z bosons each decay to pairs of charged leptons, specifically an electron and an antielectron or a muon and an antimuon, then this is also a striking signature of an unusual event, and we can measure the four leptons precisely and reconstruct the candidate 'Higgs' mass.

That's all well and good, except . . . The hint we were seeing corresponded to a Higgs mass of 125 GeV. Remember, mass is just the energy at rest. And the mass of a Z boson is 91.1876 ± 0.0021 GeV. So to make two real Z bosons we need 182.38 GeV of energy, and the Higgs just doesn't have enough. It is short by 57 GeV or so.

The key to this conundrum is the word 'real' in the previous sentence. A 'real' particle is one that lives for a long time on the timescale of our experiment. The Z bosons are not, in this sense, real. They decay after about 10^{-23} seconds (0.01 trillionths of a nanosecond), and only betray their presence via the bump in the mass distribution of pairs of leptons. It is the leptons that are real.[110]

---

[110] See Glossary: Feynman Diagrams (pp.145–50).

Here's the main diagram for this process involving a Higgs boson:

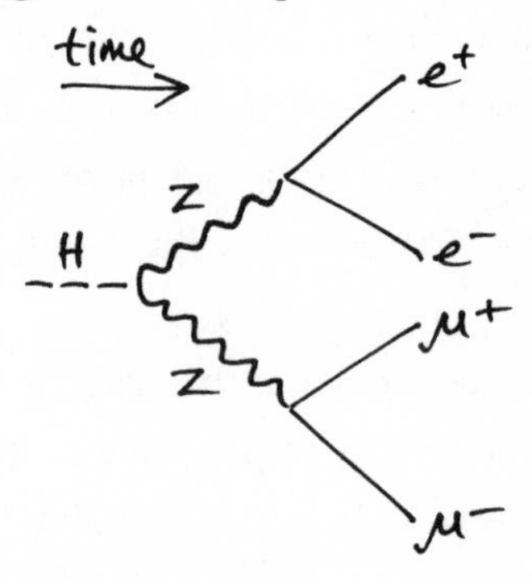

Feynman diagram of Higgs decay to two Z bosons, which then decay to four leptons

In a Feynman diagram, only the incoming and outgoing particles are 'real'. All the internal lines, including the Z bosons in this case, and the Higgs, and anything going round the loop in the previous diagram, are 'virtual' particles. They have an effect, their contribution can be inferred by measuring the real particles, but it can't be pinned down uniquely. And importantly in this context, they are therefore less restricted. Specifically, they do not have to have exactly the right mass.

Consider this plot, which summarises a lot of what we know about the Z boson:

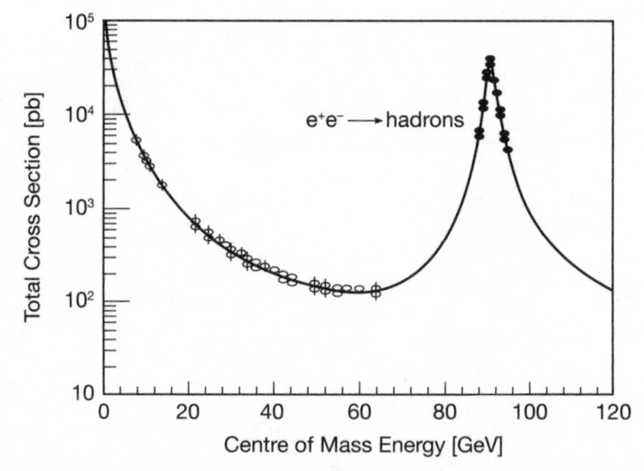

Data from CESR, DORIS, PEP, PETRA, TRISTAN and LEP machines (in Cornell, USA), DESY (Germany), KEK (Japan), SLAC (USA) and CERN (Switzerland)

A bump can be a boson, remember.[111] This bump is the Z boson, made in various electron–positron colliders up to and including the LEP2 machine running until the year 2000 in the tunnel at CERN now occupied by the LHC. Along the horizontal axis is the total energy of the colliding electron pairs; this is equivalent to the mass of the virtual particle they produce when they annihilate each other, labelled Z/γ in this diagram:

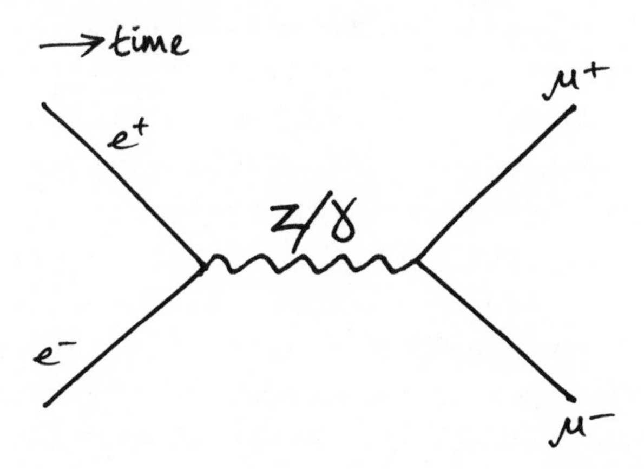

Up the left-hand side of the plot is the cross section, which is just a measure of the number of times such a collision happens for a given luminosity of incoming electrons and positrons. There is a bump at an energy of about 91 GeV, meaning 91 GeV/c^s of mass.[112] This is due to the Z boson. In a sense it is the Z boson.

You can see a lot of physics in this plot. The dominant ways in which electrons and positrons interact are summarised in that cartoon – they either annihilate to a photon (γ) or a Z. A real photon has a mass of zero. If a virtual particle in a Feynman diagram can have the correct mass, the chances of that process happening are hugely enhanced. Thus at the left of the plot, near a centre-of-mass energy of zero, the chance of a collision (the total cross section) is very high. To the right, as the energy goes up, the photon is forced further and further away from its correct mass and the cross section falls rapidly. Until, at about 60 GeV, the cross section

---

[111] See 5.1 Why Would a Bump Be a Boson?

[112] . . . if you aren't happy with natural units where $c = 1$.

stops falling, and starts to rise again. This is because the Z boson can now be produced close to its correct mass. This enhances the probabiltity of the collision, and causes the big bump at 91 GeV. Then, as the energy goes still higher, the virtual particle in the middle of the diagram has to have a mass even higher than the Z boson mass, and so the cross section falls again.

Higgs boson decays into two Z bosons are still important, even though the Higgs mass is less than twice the Z mass. The reason is in that plot. Imagine one Z boson is produced at the peak, at the correct mass of 91 GeV. The other one then has to be at about 34 GeV to make the energy sum add up. But the probability of a lepton pair being produced with this mass is far from zero in that plot, and it is similarly not zero in the LHC either. In this plot, at 34 GeV it looks like it would mostly be photons in fact, but of course we know the Higgs does not decay directly to photons. However, it will to the Z, because it is heavy, and there are some 'virtual' Z bosons being produced even at 34 GeV. The key point is that the Z peak is not just a spike at 91 GeV. It has a width, and this allows a 125 GeV Higgs to decay to two Zs. It's just that one of them will not be at its preferred mass. So even for a Higgs that does not have enough mass to create two real Z bosons, the decay to two Z bosons remains important.

Given this kind of craziness, it is important to be aware of what we really measure. That is, the leptons. It is actually impossible to say whether a given electron–positron pair came from a photon or from a Z. In fact, as I mentioned in the glossary about them, it is easy to get carried away with these cartoons, these Feynman diagrams. They are super intuitive and elegant representations of what goes on, but they are not like a time-lapse of a snooker shot. They do not represent a unique history of what happened in a given collision.

This is one of the deeply strange things about quantum mechanics, about quantum field theory, the bedrock on which the Standard Model is built. Particle physics is wrongly named. Or at least, particles aren't what we usually think they are.

Feynman diagrams represent amplitudes, not probabilities. If you understand that, you've understood something very deep about the way the world works. And interestingly, though in principle all particle physicists (despite the misnomer) know it, many of them forget it in their everyday work, and this can be dangerous, scientifically at least.

To break down that statement about amplitudes and probabilities, it's best to start where Richard Feynman started when illustrating the weirdness of quantum mechanics – the two-slit experiment. If you send waves (water, sound, light, whatever) towards a barrier that has two slits in it, and if the width of the slits and the distance between them is not too far from the wavelength of the wave, you'll see a pattern in the waves after they have passed through the slits in your screen. If they are water waves, for instance, some areas of the water will be very still, others will have quite large waves.

This is a phenomenon called 'interference', and it is to do with what an amplitude is. A wave is something oscillating about an average – the maximum distance from the average value of the something that a wave reaches. So if the peak of a wave is 10m above sea level, the average is zero (sea level), and the amplitude is 10m. The same goes for the minimum. The trough will be 10m below sea level, or at -10m.

The interference effect comes when waves get out of step with each other. If there are two sets of waves, one passing through each slit in the screen, it is possible that for some bits of sea the waves from both slits arrive at the same time. One will be peaking at 10m at the same as the other, and you get a wave with a 10 + 10 = 20m peak – they add up. When the troughs arrive they will also add up, so you get -10 + -10 = -20. This is a region of constructive interference.[113] Conversely, in some areas the trough of one wave arrives at the same time as the peak of another. In this case, they cancel each other out: 10 – 10 = 0. Everything just sits still at sea level – this is destructive interference. The amplitudes of the waves are 10m everywhere, but the disturbance in the sea is the sum of all the amplitudes, and varies from 20m to zero depending upon how they combine.

[113] Though if you are in a boat, I guess it might be quite destructive.

As with waves, so with quantum fields. The very odd thing about the two-slit experiment is it doesn't just work with water or sound waves, or light waves (which we already know are actually photons, which behave like particles sometimes). It works with electrons. If you fire electrons at a screen, you can detect them, one at a time, bing, bing, arriving at a detector, just like respectable particles (which I have to think of as very small snooker balls). But they will build up a pattern just like the waves – areas with no electrons, and areas with lots of electrons. Interference is happening.

The probability that an electron will appear somewhere is not proportional to the amplitude of the individual waves, it is proportional to the sum of them. Actually, it is proportional to the sum of them squared, so it is always positive (-20 squared is -20 x -20 = 400, the same as 20 squared). This is how Feynman diagrams work. You have to calculate all the possible diagrams that could produce a particular set of particles (say the four leptons in one of our collisions). You have to add them all up and square the result to get the cross section.[114] Sometimes adding another possibility can reduce the cross section, because it destructively interferes. Sometimes it increases it. And for any given measured set of particles you cannot say which exact Feynman diagram it was produced by, because all the possibilities contribute. Sometimes positively, sometimes negatively.

This is not to say that we can't learn anything from the measurements – the bumps in the distributions are equivalent to the patterns in the waves in the two-slit experiment – we can work out where the slits were. (In fact, such patterns in X-ray crystallography are used to work out the structure of molecules. Rosalind Franklin's patterns were used to work out the structure of DNA.) And we can work out whether there was a Higgs boson, or a Z, in some of the diagrams involved. But we can't work out exactly which diagram matches which event. All we can really measure are the final particles, the ones that live long enough to travel away from

---

[114] See Glossary: Feynman Diagrams (pp.145–50).

the collision and make the patterns in our detector. It is actually astonishing how many professional particle physicists forget this and take the cartoons far too seriously.

So don't be like them. Remember that we look for sets of four leptons in three possible combinations,[115] we plot the total mass reconstructed from them, and we look for a bump. Just like we do with the pairs of photons.

6.5 Muons: The Last Onion Layer

Muons aren't very common in everyday life. At least, they are a lot less common than photons, electrons and hadrons. Therefore if you spot one passing through your detector, the chances that something interesting just happened are relatively high. In fact, muons are crucial in the Higgs-to-ZZ search – the best sensitivity in this channel comes from the four-muon final state, when both Z bosons decay to muon–antimuon pairs. They are also vital in measuring Z and W bosons in general, and useful in searching for lots of beyond-the-Standard-Model physics.

You might think muons would be easy. Travelling from the collision point at the heart of a detector, by the time a particle has got to the muon tracking systems, the final layer of technology surrounding the collisions, it has passed through the inner tracking detectors and the calorimeter. All charged particles, including muons, should have been measured in the inner tracker, and all particles except for muons (and neutrinos, which bother no one) should have stopped in the calorimeters. Life should be quiet and simple.

In some ways that is true. But there are other challenges, and as muons are very important, the muon detectors are a high priority for any experiment. It cannot be a coincidence that the 'M' in CMS stands for

[115] Electron–positron–muon–antimuon, electron–positron–electron–positron, and muon–antimuon–muon–antimuon

'muon', and the 'T' in ATLAS stands for 'toroidal', after the toroidal magnets of the ATLAS muon system. OK, it is true that being on the outside of both detectors the muon system is the biggest component, and the first thing you see, but that just goes to show, doesn't it?

One challenge is the fact that even though the detectors are deep underground, there are still particles from cosmic rays and from background radiation that can hit the outside of ATLAS, where the muon detectors are. Possibly the major challenge, though, is the sheer surface area you have to cover to surround the inner trackers and the calorimeter. While it would be technically possible to do this with semiconductors, as is done with the inner layers of particle trackers, it would be prohibitively expensive and largely pointless, since most of the silicon would see nothing most of the time.

What is needed is a precise, fairly fast technology that can cover a very large volume cost-effectively. In ATLAS most of the muon system uses 'Monitored Drift Tubes', which are essentially long, electrically-grounded, metal tubes, about 3cm in diameter, with a positively charged wire running along the middle. There is gas in the tubes and when a muon passes by it can knock electrons off the gas molecules, ionising them. The electrons and ions drift towards the anode and cathode respectively, and a pulse is read out, from which the position of the muon can be determined. In some parts of the muon system a different layout (Cathode Strip Chambers) is used, but the principle – gas that gets ionised and a voltage to accelerate the electrons and ions – underlies it. The CMS muon system likewise uses a mix of technologies that exploit the same principle. And both ATLAS and CMS have large magnets, bending the charged muons (remember the solenoid surrounding the inner tracker was left behind long ago) so that their momentum can be independently measured.

A high-momentum muon track, measured in the inner trackers, penetrating the calorimeter, and matched to a track in the muon system, is one of the clearest and most precisely measured signs that an interesting collision occurred in the LHC. And as I said, in the crucial Higgs-to-four-lepton search, muons were a determining factor.

6.6 What Is It?

Given the trickiness of quantum field theory, where one has to state carefully what has actually been seen and then discuss possible interpretations, it was clear that if we did see something new, there would be much repeating of the questions, 'What is it?' and in particular, 'Is it the Standard Model Higgs boson?' And there would be a lot of cagey answers.

In the end, proving something exactly with experiment is impossible. 'Exactly' is a statement that requires infinite precision and if you want that, you'd better be a mathematician, not a scientist. You can never know exactly how long a piece of string is. The scientific question is really, 'Are its properties consistent with it being a Higgs boson or not?' and you have to decide how many different things need to be consistent, and to what precision, before you start calling it a Higgs. This is a matter of judgement.

The Standard Model is remarkably predictive when it comes to the Higgs boson. It predicts that it exists, for a start. And it must have zero electrical charge, and zero 'spin' (that is, it carries no intrinsic angular momentum – it is a scalar boson). The Standard Model does not predict its mass very well, but once the mass is fixed, everything else follows, including a precise prediction of what particles it will produce when it decays. The whole point of the exercise is to give particles mass, and this means that generally, the heavier a particle is, the more likely it is to be produced in a Higgs decay, assuming it is possible to do this and conserve energy.

In 2011, of course, we had no significant signal. However, we did have sensitivity, such that the data *could* have been inconsistent with the existence of a Higgs boson, down to about 130 GeV in Higgs masses. But the data remained consistent. This statement relied mainly on comparisons to predicted event rates. We also knew something about the distribution of the events, with a possible but not solid mass peak.

To start calling this a 'Higgs boson' rather than an 'excess' or a 'candidate', we would not only need to have much more confidence that it was real, we would have to measure more of its properties. The main thing would be to see the whatever-it-was (assuming it was something) decay in at least two, and preferably more, different ways. If the relative decay rates were to look as expected for a Higgs boson, it would become pretty compelling, progressively more so as more decay modes were accumulated. The more we measured, the more the discussion would move from 'possible excess' to 'Higgs-boson candidate' to 'Higgs boson' to 'Standard Model Higgs boson'.

And even if we did find something, we would never be able to show a picture of an event and say, 'This event definitely involved a Higgs boson.' This is a bit of a shame, but the reason behind it provides a resolution to a severe parental dilemma, and explains why I am in fact sometimes the tooth fairy. Bear with me.

One of the developing pieces of evidence for a Higgs was the tiny bump in the two-photon mass distribution. I can certainly show a collision event containing a pair of photons that exactly gives the 'Higgs mass', i.e. at the top of the bump. But it would still not be possible to be sure that a particular pair of photons came from a Higgs boson. Very roughly, if at the peak there were about 70 events, probably only about ten of these would be due to the new boson. With our detector we can't tell the difference between these and the 60 or so background events. In fact, even with a perfect detector, some of that background would remain and be quantum-mechanically mixed with the signal. All that is definite is what goes in and what goes out. These you can measure.[116]

There may be several possible ways of producing a set of new particles from the incoming ones, but if the resulting set is identical, it is not physically meaningful to say which possibility occurred. To calculate the probability of that result occurring, you have to add up all the possibilities in a particular fashion.

Now to the parental dilemma. It is especially acute around Christmas,

---

[116] The actual numbers of Higgs candidate events are much higher than this, but the principle is the same.

but if you have children who are losing their milk teeth, it is ever present. Is Father Christmas real? What about the tooth fairy?

Do you spoil the fun or do you lie? Something in me hates the idea of lying to my kids and undermining trust. On the other hand, I don't want to be a miserable bastard. Here's my way out.

Anything that has the same initial state (tooth) and final state (money) might in fact be an event in which a tooth fairy was present. To put it another way, anything that removes the tooth and delivers money shares such an essential property with a tooth fairy that it can be said in a sense to be one.[117]

These days, my son doesn't believe a word of it, of course. But in the early days it was *the truth*. We managed this transition without lies, betrayal or tears because actually, when tiptoeing into the bedroom with a shiny pound coin, I really am the tooth fairy. I am, of course, at the same time Dad. This seemed to work, and now he's older, it's still fun.

It's not much of a stretch to extend this to Father Christmas, and it also explains why sometimes Father Christmas uses the same wrapping paper as your parents – he and they are, in a sense, indistinguishable quantum possibilities for the delivery process.

Maybe. Either way, in physics the answer to 'What is it?' is always, 'It's a thing that behaves like this.' And we were feeling increasingly confident that we were closing in on something that behaved very much like the Higgs.

---

[117] Anything removing both teeth and money is probably a dentist. Or possibly a mugger.

SEVEN

Closing In

January–June 2012

7.1 Eight TeV

Obviously, after these tantalising hints the pressure to follow up and answer one way or the other was going to be huge. The BBC showed up, filming a *Horizon* programme that started off being about particle physics in general, but grew increasingly focused on the Higgs hunt as it became clear things were heading for some kind of conclusion.

On 7 February 2012, the Higgs search results from December were submitted to journals and the arXiv.[118] The break between this and the talks in December had involved a lot of cross-checking and a lot of work (and Christmas), but had not seen a great deal of change in the results. After peer review, the ATLAS and CMS papers were both published, essentially providing retrospective vindication of what we had shown.

On the same day these papers came out, I went to Google London (the offices, I don't mean I searched for it, I know where London is) for breakfast and the launch of a film about CERN. Two nice things about this. I met Phillip Greenish, one-time member of the council of STFC, and CEO of the Royal Academy of Engineering at the point that it had

---

[118] http://arxiv.org/abs/1202.1408; http://arxiv.org/abs/1201.1487.

suggested (during a crucial part of a UK funding debate) that particle physics should bear the brunt of the cuts. His presence at what was an LHC-based launch, and some of the things he said, seemed to imply that perhaps we'd moved on a bit from those fratricidal attacks. Also, in the film I liked a snippet featuring US colleague Zach Marshall at the ATLAS barbecue. He talked about theorists inventing weird new physics scenarios that might show up at the Large Hadron Collider:

> You have these people trying to just predict, and predict as many different things as they can. Because if one of them is right, they'll be famous. And if all of them are wrong – they'll be like everybody else!

And he's right. So there you have it. Experimentalists get ignored if they are right (e.g. about the speed of neutrinos), and hugely cited if they are wrong. Theorists are ignored if they are wrong, but get a Nobel Prize if they are right.

Not quite true, but not completely false, either.

The LHC started giving us data again on 5 April. Preparing for the new collisions required quite some work, because they were at a centre-of-mass energy of 8000 GeV (so beam energies of 4000 GeV each), an increase from the 7000 GeV of 2011. This was generally a good thing – more energy means we have more chance of creating new particles, for instance, since the heaviest particle you can possibly produce has a mass of $m = E/c^2$ where E is energy and c is the speed of light (exercise for the reader – rearrange that to make a famous equation). But a change of energy did also mean we had to revisit all our simulation programs and get ready to check that they would work just as well at the new energy as they did at the old.

At a lecture I gave in February to 16–19-year-olds as part of the Institute of Physics' 'Physics in Perspective' series, I was asked how come, if we have proton–proton centre-of-mass energies of 7000 GeV, the reach in energy for new particles (including the Higgs) is much lower than that?

In the specific case of the Higgs (where most of the action was by now down below 130 GeV), there is plenty of energy, but the difficulty is more to do with identifying a Higgs boson amongst the backgrounds. However, there's a more general point. ATLAS and CMS can make the equivalent of the plot shown in section 6.4. In this case, instead of electrons and antielectrons annihilating to a photon or Z boson, which then decays, the incoming particles are quarks and antiquarks, like this:

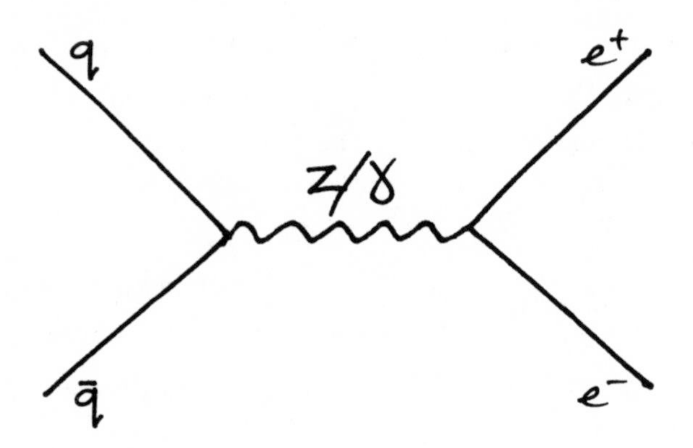

As you already know,[119] there are plenty of antiquarks, and gluons, inside our protons, as well as the quarks. So that is where the antiquark comes from. In the ATLAS and CMS plots, there is also a bump at 91 GeV due to the Z boson, just as in the electron–positron plot. But the LHC gets us to much higher energies than LEP – as far as about 1500 GeV, compared to about 100 GeV with LEP. In its later running as LEP2, the machine went up to about 210 GeV, but that was the limit. That was the highest collision energy the beams could get to. However, even though the LHC was delivering collision energies of 8000 GeV in the search for bumps, it doesn't get close to this hypothetical limit. Why is this?

The reason is, of course, that protons are not fundamental particles. The really short-distance, energy-frontier physics takes place at distances much smaller than the proton radius, so effectively, from this point of

---

[119] From 4.5 Inside a Proton.

view, the LHC is a quark and gluon collider, rather than a hadron collider. And unfortunately, even though the protons have an energy of 4000 GeV each (so a total energy of 8000 GeV available), any given quark or gluon only carries a fraction of the full energy of the proton, so the available energy to make new particles is generally a factor of five or ten lower than the proton energy might indicate.

7.2 The Curse of Meetings

The summer conferences were clearly going to be the make-or-break moment in the Higgs search. It hardly seemed possible, but the frequency of meetings intensified.

Given the time zones involved, it would be possible to spend every hour of every European working day, and most of the night, in an ATLAS meeting. Since they are nearly all available via some form of teleconference, with enough connections you could spend most of the day in half a dozen of them at the same time. This would of course melt your brain.

To add insult to injury, a curious phenomenon has emerged. The moment a meeting begins to get interesting, one of the participants (usually the chair) will almost invariably suggest they 'take it offline'. And we move on to the next topic.

There are a few cases which provoke this.

1. The unexpected moron

Someone, perhaps senior (sometimes me), has totally missed the point of the presentation or discussion. They ask a question displaying such profound ignorance and confusion that the rest of the meeting can only stare at their shoes or the ceiling in embarrassment. A variation on this case is simply a question so basic ('What particles are we actually colliding?') that one can only assume the questioner walked in off the street by mistake. Either way, a good chair moves that discussion offline as soon as possible to save everybody's blushes.

2. *The grudge match*

There aren't so many particle-physics experiments in the world. If, as a PhD student, you meet an arrogant pedant suffering from acute Dunning–Kruger effect,[120] the chances are you will bump into them, on and off, for the rest of your career. If they ask you a stupid, tedious and hostile question in a meeting, it might provide a diverting floor show for the other participants, but the chair will most likely take it offline. You should agree to this. Perhaps take it to a darkened corridor, with a couple of menacing friends. 'Ahem. Jon wanted us to talk to you about that question you asked . . .'

3. *The technical tumble*

To be honest, this one is where some actual progress is made. Someone in the meeting has deep technical knowledge of an issue that has just been raised. It is pertinent to the topic, and indeed quite often someone presenting a result will realise immediately that this needs checking in detail and would rather go into a huddle and do this than potentially crash and burn in front of their colleagues. Also it may take days. Take it offline.

4. *The dodgy connection*

Someone suddenly starts talking as if they have inhaled helium, turned into a Cyberman (for Doctor Who fans) or put their head into a bucket of water. This is always blamed on a poor Internet connection or bad microphones, though I suspect in at least a couple of cases Cybermen have really been involved. While it is undoubtedly true that the teleconferencing equipment is generally to blame, this is also a handy cover if you find yourself unwillingly embroiled in any of the examples above. It more or less forces it offline.

---

[120] Delusions of superior ability. Pretty much the opposite of the much more common 'impostor syndrome'.

Oddly, in almost all cases the 'take it offline' phrase means 'continue it online in a series of lengthy emails'. So perhaps we aren't so cool after all.

There are many good things about Switzerland. The flag is a big plus. Also – possibly in reaction to the excess of meetings at CERN, the United Nations or the World Trade Organization – a Swiss political party formed in 2011 with a policy to ban PowerPoint. Contingency plans, including holding all meetings over the French border in the Prévessin site of CERN or (clearly very popular) enforcing a strict LaTeX-only[121] policy, were discussed extensively.

In meetings.

7.3 Waves

Angels and Demons is a best-selling thriller by Dan Brown, and a film starring Tom Hanks. It features CERN quite heavily. When people ask me about it, they expect to be disabused.

'Yeah, yeah, of course we don't have private jets, or parachute-training-tower whatsits,' I say.

'I don't even own a white coat. It's fiction, enjoy it!' I say.

'What's that? Antimatter? Oh no, that's real, obviously.'

In fact, antielectrons (i.e. positrons), antiprotons and other antiparticles are commonplace in even moderate-energy physics facilities. Even in hospitals, in positron emission tomography (PET). This is a diagnostic technique in which a radioactive isotope of an element that decays and emits positrons is introduced into the body.

An isotope is an atom in which the atomic nucleus has the same number of protons as usual, but a different number of neutrons. The radioactive decay time of an atomic nucleus is fixed by the mix of neutrons and protons it contains. However, chemical properties of an element

---

[121] No, not a kinky dress code. LaTeX is a typesetting program, written largely by scientists, which I wish I had written this book in, really.

depend only on the number of electrons, which in turn is fixed by the number of protons, so that the electric charges cancel out and the resulting atom is neutral. Thus the unstable isotope carbon-11 with six protons but only five neutrons will be chemically identical to the stable isotope carbon-12, which has six of each.

This means, for example, that molecules and compounds such as sugars or proteins can be made containing carbon-11. These sugars or proteins will behave just like the common versions, but the carbon-11 nuclei will decay after some time, emitting a positron.

Hanging around inside a human body, this positron will quite rapidly meet up with an electron and they will annihilate, producing two photons. Each photon will carry an energy very close to the mass of an electron multiplied by the speed of light squared (it's that equation again), which in particle-physics units is 0.51 MeV. By measuring these photons, it is possible to map out very precisely (and for a very low dosage of radiation) where the isotope went in the body and this, for a careful choice of isotopes and molecules, can tell you rather a lot about what is going on in there.

Antiprotons are harder to make, simply because they are nearly 2000 times heavier and so it requires about 2000 times the energy to create one. Even so, by particle-physics standards this is very doable. The problem is, they have to be created in a collision with something, usually a proton beam hitting some material, and when they are created, most of them are moving quite quickly. The tricky bit is to catch them, slow them down, store them and – if this is your plan – let them combine with positrons to make antihydrogen.

All the while, you have to stop them from annihilating with all the normal matter hanging around. Then, if you want to really know what you have got, you have to store them long enough to study their properties.

The ALPHA experiment at CERN did all this, and published the results in March 2012. Not only did they trap atoms of antihydrogen, but they fired photons (microwaves in this case) at them. These photons make the positrons inside the antihydrogen change energy levels. They can make the spin of the positron change its alignment, so that it either lines up with, or against, the spin of the proton. When this happens, the magnetic properties of the antiatom change, which means it isn't trapped any more, it escapes and then annihilates.

The frequency to which the microwave has to be tuned in order to cause this gives a measure of the difference between the various energy levels inside the antihydrogen. To within the precision of the measurement, the difference between them is the same as it is for hydrogen.

This was the first time anyone had measured inside antiatoms. The result – 'It looks the same' – was not a surprise. It would have been a huge surprise if it had been different. The symmetry between matter and antimatter in their interactions with photons is built very deeply into our theory. The existence of antimatter was predicted by combining quantum theory and special relativity in the Dirac equation, and within this theory the electric charge of the electron has to be *exactly* opposite that of the positron, and their masses have to be identical. The point is that even if the initial measurements weren't very precise, and even if theory told us the answer to expect, this was the first actual measurement. More precise ones will come. Prediction and extrapolation from theory, even a very well-tested theory, are not the same kind of knowledge as an actual measurement.

The field of antimatter spectroscopy had just begun. Matter spectroscopy, on the other hand, has been around for a long time and is a stunning branch of science with enormous implications and applications. It is through spectroscopy that we know what stars and galaxies are made of, and the presence of different distinct emission or absorption lines in the light from particular elements was one of the main puzzles driving the development of quantum mechanics.

Because they are fermions,[122] the electrons around an atom cannot all be in the same quantum state, which means they can't all be at the same energy. An even deeper property, though, is the fact that they are in

---

[122] See Glossary: Bosons and Fermions (pp.31–3).

'quantum states' at all. What this means is that there are only certain values of energy that electrons are allowed to have when they are bound to an atomic nucleus. Because of this, if you want to move an electron from a low-energy state to a higher-energy state you have to put in exactly the right amount of energy. The energy would normally come in as a photon. A fixed energy for a photon means a fixed wavelength, so a fixed colour, if it is in the visible part of the spectrum. And that is the heart of spectroscopy.

If you have a whole bunch of atoms at a high temperature, for instance in a star, this means they are continually bumping into each other, swapping photons, and the electrons around them keep changing energy levels. Imagine two specific energy levels, for instance in sodium[123] in a street lamp. The energy gap between them has a fixed value, corresponding to a fixed wavelength of yellow light in this case. If you separate the light from a sodium lamp into its different wavelengths (as Newton did for sunlight using a prism), you'll see a bright band corresponding to electrons jumping down from the higher-energy level to the lower, and emitting a photon as they do so.

The element helium was discovered like this, in sunlight, during an eclipse in 1868. A French astronomer, Pierre Jules César Janssen, spotted the bright line at 587.49 nanometres, corresponding to an energy of 2.11 eV, and coincidentally also appearing as yellow, quite near the sodium lines. A couple of months later, Norman Lockyer also spotted the line and correctly concluded that it must come from an element not at that point known on Earth, but present in the Sun. It was named helium (from the Greek *helios*, meaning the Sun) and first detected on Earth by the Italian Luigi Palmieri 14 years later.

Spectroscopic analysis of the lines in light – sometimes 'emission lines' like these, where there are more photons because of electrons jumping down a level, and sometimes absorption lines, where there is a darker patch because electrons are absorbing photons at a certain wavelength to

---

[123] The example I used already in section 2.1.

jump up a level – can, in laboratories on Earth, be used to spot tiny traces of elements or compounds in samples of material. They can also tell us the composition of objects in space we could never reach to study directly. Astonishingly, the fact that these patterns of lines shift lower in energy (and therefore longer in wavelength, towards the red side of the spectrum – 'red shift') when something is moving away from the observer is what allowed us to see that the universe is expanding. All the distant galaxies have the characteristic patterns of spectral lines shifted to longer wavelengths, so they are moving away from us.

Spectroscopy gets us a long way, but it is just one aspect of the concept of waves in physics. Waves are possibly the best bit of physics. They pop up everywhere. The fact that an electron behaves like a wave is why distinct energy levels exist around atoms in the first place. Niels Bohr put together the first model of atoms that could explain the lines in spectroscopy, by proposing that the electrons orbited the nucleus in certain allowed orbitals, with the ones in between them forbidden.

If you imagine the electron as a wave, this provides a really good way of seeing why some orbits might be allowed. They will be the orbits where a whole number of wavelengths just fit with the length of the orbit, so that when the wave goes round the orbit, it meets itself at just the right point to add up, and you get a 'standing wave'. Any wave that doesn't meet itself at the right point, so that the peaks line up with each other, will cancel itself out, and the energy corresponding to that wavelength is therefore not allowed.

The business about standing waves may sound a little esoteric, but not only is it very close to the eventual quantum mechanical picture that emerged from solving Schrödinger's equation for electrons around atoms, it is also, on a bigger scale, how musical instruments work. The string of a harp gives out one principle note, corresponding to a vibration in the string with a wavelength that is double the length of the string. What is happening is that waves move out in both directions from the point at which the string is plucked, then reflect off each end of the string and add up (interfere) with each other. Just like Bohr's electrons, if the peaks coincide, the wave survives, and if not, they cancel each other out. Thus the string plays a particular note.

At a concert in the Barbican once I found my mind drifting to physics. This doesn't imply any criticism of the music, which was stunning, but the fact is that when stunned, my mind often wanders off. It considers the fact that the London Symphony Orchestra logo does in fact look like a conductor with a baton. It admires Susanna's cheekbones. It wonders what Szymanowski did with the rest of his vowels, and what it must have felt like to be Bartók, creating beauty while Europe headed towards horror. And in the middle of Bartók's Violin Concerto no. 2, it began wondering about physics and harps.

The orchestra had two harps. Beautiful, sweeping things, nearly triangular but with a distinctive curve in the top edge. Why is that curve like that? I wondered. Maybe because it looks nice, but since the curve affects the length of the strings, and the length of the strings affects the note played, it was probably not just ornamental.

Four quantities characterise a wave: speed, frequency, wavelength and amplitude. You could add a fifth maybe, the shape of the wave, which is where all the tone and other subtleties come in. That's where Stradivarius made his cash. But being a physicist I want to keep things simple.

Amplitude is a bit boring. It is just the height of the peaks of the wave, or the depths of the troughs. So for sound it is basically the pressure difference between the highest-pressure bit of a wave and the average pressure. It's the volume.[124]

Speed, frequency and wavelength are related: speed is equal to the wavelength multiplied by the frequency. And in fact the speed is a property of the medium through which the wave travels. The speed of sound in air (at room temperature and pressure) is a fixed 1,236km per hour. So for a sound wave, once you have fixed the frequency, you have also fixed the wavelength.

---

[124] I'm not sure why we use the word volume for loudness, really, but at least amplifiers do increase amplitude, so that's nice.

The frequency is what we hear as the pitch of the note. When a harp string is plucked, it vibrates with a certain frequency, vibrating the soundbox and compressing and decompressing the air inside it, thus making sound waves of the same frequency and, in the right hands, leading to music.

The frequency of the vibration in the string is set by the length of the string, the tension in the string and the material it is made of. If you want to have all your strings made of the same stuff (so they have similar tone) and at the same tension (so they take the same effort to pluck), you have to increase the length to get deeper notes. Unfortunately, you have to double the length every octave. This means an exponential growth in length.

A triangular-shaped harp, without the curve in the upper frame, gives only a linear growth in string length, not an exponential one. If you wanted to cover enough octaves, this would lead to an unfeasibly big harp. The curve in the top of the frame looks to me as though it is there to allow the lengths of the strings to grow exponentially as far as this is feasible. So, for the shortest strings, they can have the same tension, be made of the same stuff and be about equally spaced along the frame. But if you extrapolate that curving shape as you go away from the harpist, you can see the harp would get too big before you got to the next octave. So the curvature in the frame changes, and the type of string has to change, too, to keep the frequency dropping exponentially and thus allowing more octaves. So the graceful curves are functional as well as ornamental.

I often use harp, or double-bass, strings as an analogy to try and explain one of the reasons why high-energy collisions are an exciting place to do physics. One way of explaining it is that since $E = mc^2$, we need a lot of E to make a new particle with mass, m. Another is to talk about how we are probing the physics of the early universe, a few moments after the big bang.[125] But my favourite is to talk about waves and resolution.

---

[125] See 4.6 Heavy Ions for Christmas.

Which gets us to the question: Is light a wave or a particle? What about an electron? A gluon?

The answer I give myself to that whole wave/particle duality question is, 'No.' Or rather, I tell myself that the concept of a wave, like the concept of a particle, is an imperfect everyday analogy for a quantum of something. So, 'particle' describes some of the ways a photon, or an electron, behaves, and 'wave' describes others. In the end the reality is an excitation in the quantum field.[126]

Certainly, quanta have energy, momentum, wavelength and frequency, and all these are related to each other. Energy and frequency are proportional to each other, and momentum is proportional to the inverse wavelength.[127] So high energy means high frequency. And high momentum means short wavelength. If you are talking about a high-energy collider like the LHC, the momentum and energy are proportional to each other as well. So the bottom line is that high energy means high momentum means short wavelength.

This is important because the wavelength sets the resolution, or, if you prefer, the size of the smallest thing you can see. Radar, with a wavelength of metres, is great for seeing boats and planes. But if our eyes were sensitive only to radar, we would not be able to see each other, because we are smaller than the wavelength. Optical light, with wavelengths of hundreds of nanometres, is much more practical and allows us to see rather fine detail. It is also lucky that this is the peak wavelength of the Sun's radiation, of course. Given that our eyes presumably evolved to make use of the brightest wavelength, if the Sun mainly radiated in the infrared, we'd have infrared-sensitive eyes and everything would start getting blurred just below a millimetre. Tricky for threading needles.

Because it has the highest-energy beams ever achieved in a laboratory,

---

[126] See Glossary: Fields, Quantum and Otherwise (pp. 57–60).

[127] In the appropriate units, energy (E) equals Planck's constant multiplied by frequency (n), and momentum (p) equals Planck's constant divided by wavelength (l). Also, for a massless particle, $E = pc$ and $nl = c$. I put this here in case it helps. Equations seem clearer than words sometimes. But don't worry if not.

the LHC therefore has the shortest wavelength quanta ever (mostly gluons, in fact) and can therefore probe nature at the shortest distances we have ever been able to reach. We are looking not only into the heart of an atomic nucleus, but into the protons inside it, and at the quarks inside that. And potentially, if they have anything inside them, we might see that too. This is the way I prefer to think about the 'energy frontier'. And this is why we learn about small things like quarks from big things like the Large Hadron Collider. It is also because – even at these tiny distance scales – quarks and electrons do not appear to be made of anything else that the puzzle of mass shows up here. This is where either we find out that electrons and quarks are not fundamental and are made of some other stuff – or we find a Higgs boson to give them mass.

7.4 The Neutrino Matrix

One of the weirder things about QCD and the weak force is that they are non-Abelian[128] theories. This means that the algebra of the symmetries at the heart of the force[129] does not commute. Which means that A times B does not equal B times A.

Now common sense, and experimenting with counters, will tell you that for two numbers, A and B, multiplying A by B always gives the same result as multiplying B by A. Two lots of three pence make six pence, and so do three lots of two pence. This is always true for numbers, in fact. However, it is perfectly possible to define kinds of mathematics where AB is not equal to BA. Indeed, mathematicians have been studying such things for a long time.

What might be more surprising is that physicists use it all over the place, because there are physically relevant objects for which AB does not

---

[128] After the Norwegian mathematician Niels Henrik Abel, who was not, as far as I know, murdered by his brother.

[129] See Glossary: Gauge Theories (pp.96–9).

equal BA. One way of representing such an object is by using a matrix. Matrix mechanics is included in the first-year Mathematical Methods course I was teaching at UCL around this time. Because my school did an experimental 'New Maths' syllabus, I even learned something about it when I was about 15.

In maths, a matrix is just a bunch of numbers arranged in rows and columns. When you multiply two matrices, A and B, together to get another matrix, C, you multiply rows by columns.[130]

This kind of matrix may sound less enthralling than an all-controlling virtual-reality supercomputer filmed in bullet time and black leather, but it is also more useful.

You can use a matrix to describe what happens when you move something around, for example. This is a case when order (AB or BA) clearly matters. If you first turn 90 degrees, then walk 10 metres, you obviously get to a different place than if you first walk 10 metres, then turn 90 degrees. If B is a matrix representing the turn, and A represents the walk, then the combined 'walk after turning' matrix (C = AB) must be different from the combined 'turn after walking' matrix (call it D = BA). So C doesn't equal D, AB doesn't equal BA. If AB always equalled BA, matrices would be useless for representing such operations. It's because they don't commute – that is, they are non-Abelian – that they are useful.

Matrices also proved to be what Paul Dirac needed when he was trying to work out the quantum mechanics for electrons moving at high speeds. In fact, the features that forced him to use matrices were the same features that allowed him to describe the spin of electrons, which is crucial to the whole behaviour of atoms and the structure of the periodic table. The same features also led him to predict the existence of antimatter.

It amazes me that mathematics and the behaviour of the real world

---

[130] That is, multiplying the first row of A by the first column of B gives the entry in the (first row, first column) position of C; the first row of A by the second column of B gets you the (first row, second column) entry of C; and so on.

seem to be so closely connected. Good research is as much about choosing good questions to address as it is about answering them. There are always more questions than answers, and research costs time and money, so we have to make choices. Maths is one very good way of interrogating data and using it to suggest which new experiments are the most interesting, even when the methods and results, like matrices and antimatter, can seem rather exotic.

In that spirit, before continuing with the Higgs hunt, here is a final detour into neutrinos, this time for a real and important result. On 7 March 2012, new results[131] were announced by the Daya Bay Reactor Neutrino Experiment in China. These results have had a huge impact on the Standard Model, and on the future of particle physics. If you just want to follow the Higgs story, feel free to skip lightly ahead, see you in the next section. Neutrino fans, hang on.

Neutrinos mix up amongst themselves,[132] and this mixing is a sort of rotation between one set of neutrino labels (the flavour – electron, muon or tau) and another (the masses). This rotation can also be described by a matrix, containing three mixing angles.

A vital open question was whether the three types of neutrinos genuinely mix three ways, or whether what we were seeing was just two separate pair-wise mixings. Two of the three possible mixing angles had already been measured, but one remained that had not.[133] It is called θ_{13} (theta one-three).

Daya Bay produced[134] the first clear measurement of θ_{13}, showing that indeed it was not zero – in fact it is about 9 degrees. This was quickly supported by a similar measurement from the Korean RENO experiment.[135] Since θ_{13} is a fundamental parameter in the Standard Model of particle

---

[131] http://arxiv.org/abs/1203.1669.

[132] See 5.5 Meanwhile in the Neutrino Sector.

[133] Though the MINOS and T2K long-baseline experiments, as well as the Double Chooz reactor experiment in France, had some evidence that it was not zero.

[134] http://arxiv.org/abs/arXiv:1203.1669.

[135] http://arxiv.org/abs/arXiv:1204.0626.

physics, this would be an important measurement anyway. But there was a bit more to it than that.

If θ_{13} were zero, we would just have had two-way neutrino mixing. The flavour states might just mix up neutrinos 1 and 2, and neutrinos 2 and 3. With θ_{13} bigger than zero, neutrino 1 also mixes with neutrino 3. In this case, and *only* in this case, a fourth parameter is also allowed in the matrix. This fourth parameter (delta, δ) is one we haven't measured yet, but now we know it is there. And the really important thing is, if it is there, and also not zero, then it introduces an asymmetry between matter and antimatter.

This is important because currently we don't know why there is more matter than antimatter around. We also don't know why there are three copies of neutrinos (and indeed of each class of fundamental particle – three charged leptons, three charge -⅓ quarks and three charge ⅔ quarks). But we know that three copies is the minimum number that allows some difference in the way matter and antimatter experience the weak nuclear force. This is the kind of clue that sets off big klaxons in the minds of physicists: *New physics hiding somewhere here!* It strongly suggests that these two not-understood facts are connected in some bigger, better theory than the one we have.

We've already measured a matter–antimatter difference for quarks; a non-zero θ_{13} means there can be a difference for neutrinos too. More clues.

So, not only had we seen the start of antimatter spectroscopy, which will check that there really is no matter–antimatter asymmetry in the electromagnetic force. We now, in the same week, saw the result showing that experiments such as T2K in Japan and NOvA in the US, which are looking for matter–antimatter asymmetry in the weak force, amongst neutrinos, were not wasting their time.

Chinese and Korean scientists have long collaborated on major particle-physics experiments, but these are two of the biggest particle-physics results to come out of experiments actually constructed and operated there. It used to take a long time for results to cross continents, but not any more. Like most physicists (but, disappointingly, unlike the ALPHA antihydrogen experiment at CERN), the Daya Bay people put their result up on the arXiv. This has been standard practice in particle physics and astronomy for quite some years, even predating the World Wide Web. I did my doctoral research in the days before the World Wide Web (just). But we had the Internet, and when Herbi Dreiner and I had finished what was my first ever academic paper,[136] he added it to a 'bulletin board' managed by Los Alamos National Laboratory in the US. It was then emailed around the world to particle physicists who subscribed to the bulletins.

This was all new to me. Theorists were ahead of the experimentalists (as usual). But pretty soon experiments joined in. And when the web came along, the whole thing moved over there and is now the wonder that is the arXiv, curated by Cornell University, funded in an international collaborative model and free at the point of submission and at the point of access. My first paper is still there. The arXiv stores the full text and figures of papers.[137] Basically, all particle-physics, astrophysics and astronomy papers are there, regardless of whether they are also in a journal or not. There is also coverage in condensed matter, nuclear physics, mathematics, biology and more, though I don't know what fraction of publications in those areas are uploaded.

To see why this is important, consider the fact that my first paper was also published in a journal called *Nuclear Physics B*. Last time I checked, you could still get it there, too, a snip at $31.50, of which Herbi and would I receive . . . well, nothing. Unfortunately the only way of getting the ALPHA results seems to be to pay Nature Publishing Group $32, despite the fact that your taxes probably already contributed to the cost of the experiment. Prohibitive fees either at publication time or when you

---

[136] See 3.5 Supersymmetry.

[137] Though if you look at the paper (http://arxiv.org/abs/hep-ph/9211204) you might notice we didn't upload the figures for our paper. This was because I was still struggling with postscript inclusion with LaTeX. These days, figures are all there. And if you look on INSPIRE-HEP you can see that KEK even scanned it in with the figures. We posted it to them. Ah, those were the days . . .

want to read the paper restrict the results of research to an elite of well-funded institutions and individuals. They deprive scientists elsewhere of opportunities, thereby depriving science of their skills. Demand for open access to research results is growing, and we had already decided that any definitive result on the Higgs would be freely accessible, even if this ruled out some very prestigious journals.

7.5 Is Nature Natural?

Mostly I work on measuring what happens in proton–proton collisions at the LHC. But just as the 2012 LHC running began I went to a meeting about what wasn't happening – and still hasn't so far. The meeting, at the University of Maryland, was 'Supersymmetry, Exotics and Reaction to Confronting the Higgs'. SEARCH – a quality acronym (type 1) if ever there was one.

As already discussed,[138] at the energy the LHC can reach, special things happen in physics. Electroweak symmetry-breaking happens, the W and Z bosons have masses in this energy range, and that was why we were sure that if the Standard Model Higgs boson existed, it would eventually be visible at the LHC.

More general but weaker arguments based on 'naturalness' are often used to argue that other new physics should show up at the LHC too. Naturalness, in this context, is the assumption that the parameters in a theory should be about the same size as each other. So, say the ratio of two parameters should be between 0.1 and 10, and that these ratios should not have to be fantastically fine-tuned in order to make the theory work.

In the Standard Model, unfortunately, it looks like the Higgs mass does have to be fantastically fine-tuned. This is one of the problems supersymmetry might solve. For that to be the case, we ought to see some evidence for SUSY particles, and we hadn't. Many physicists were getting

---

138 See 5.8 Which Leads, Theory or Experiment?

puzzled by the absence of supersymmetry, and in fact the absence of any new physics that could do its job, in the LHC data. It was too soon to give up on the idea of naturalness, but things had gone far enough that at the end of the SEARCH meeting, several eminent theorists advised people to work on understanding the QCD. Partly because it is interesting but mainly because if we can understand it better we can search more effectively for new physics that might be hiding in the data at the LHC.

This is, of course, what lots of us are doing already. For example, at about this time ATLAS released the first measurements of some new jet substructure variables, useful for searching for new particles. The conclusion of this new ATLAS paper was that we do understand QCD (and our experiment) well enough to use these variables; some of the ideas had even been used already in the CMS Higgs search. All this was close to my heart because I'd worked on those studies myself (they were the topic of the regular 'Boost' meetings), but many similar precise measurements, and calculations, need to be done before we can learn all the LHC can tell us. If in the end we do not find supersymmetry, or something similar, the idea of naturalness, that the theory should be elegant and not require fine-tuning, will be in real trouble.

I went through Washington (Dulles) on the way to a physics meeting and queued for ages. There were about 30 desks, only one of which was manned, and the guy at that desk spent the whole time dealing with one family. I suppose the fact that the woman was wearing a niqab and the husband had a very full beard might have slowed them down a bit. In the end, the other staff got embarrassed and let us use some of the desks allocated to US citizens. This was my second-worst experience with US Immigration. My worst so far had been on my first trip to the US in December 1995, to Penn State University. That trip was made entirely to get my visa processed. It almost failed.

I'd accepted a position with Penn State University. I had only a vague idea where this was, but that hardly seemed to matter since it was a good university and wanted to pay me to live in Hamburg and do physics with the ZEUS experiment, which is what I wanted to do.

To get my visa processed (a J-1, which would allow me to work for a US institution and get paid), I actually had to go to Penn State, however briefly. I was supposed to fly out a few days after the oral examination for my doctorate. Not wishing to take anything for granted, I had not booked my plane ticket, but was otherwise ready to go. My exam would be on the Thursday and I would fly out to the US the weekend after, assuming I passed.

Unfortunately, the Saturday before, my bag got stolen. In it were some brand-new M&S underpants, a very long scarf I had knitted myself and my passport, including the J-1 visa.

The pants were easy to replace, the scarf impossible (I had forgotten how to knit). The visa and passport could be replaced but it would be tricky, and it would have to be quick.

The first half of the week was spent on trains, going to Manchester (parents, birth certificate), then Liverpool (Passport Office). I then popped back to Oxford for my exam. On the plus side, I didn't have time to get nervous. I passed, although I still wince when I remember some of the things I got wrong. A fellow DPhil student, Rick Gaitskell, then drove me to London. Friends in need, with cars, are great. Rick is a professor at Brown University these days, working on dark matter searches. I hope he has a better car now.

I rushed to Grosvenor Square and bounced off the US Embassy because I had my travel bag with me. No bags allowed. What to do? Eventually I remembered I was a member of the Institute of Physics, and sure enough those lovely people let me leave my bag with them. Back to Grosvenor Square. Visa, done. Travel agent. Ticket, done. Hurry, hurry.

On the tube, I looked at my tickets. They said Newark. Where the hell was that? I had asked for tickets to New York. The guy must have misheard! Newark could be on the West Coast for all I knew!

Eventually I stopped palpitating. I can't remember whether I asked a random tube traveller or at the airline check-in desk whether Newark was near New York, but either way I was reassured. I barely remember the flight, but I do remember being rather spaced out when my new employer, Jim Whitmore, met me out of Immigration.

I was then driven through heavy snow from Newark to State College, PA, geometrically centred in Pennsylvania (i.e. essentially nowhere). The car nearly spun off the road at least once. ('Should have come in the pickup, said Jim nonchalantly), but I was delivered, dazed, to a motel. I remember marvelling at the size of the toilet bowl, worrying that I might fall in. Later that evening I woke up in bed, completely confused as to where I was.

There's a scene in the film *10* in which Dudley Moore, having drunkenly followed Bo Derek from California, wakes up in Mexico and staggers astonished and disorientated onto the balcony to the sound of Spanish guitars. I did this, although in my case a winter wonderland of small-town America in the deep snow, rather than a row of frantic strummers, greeted me.

Stumbling down the snow-lined street like Jimmy Stewart in *It's a Wonderful Life* (America is so cinematic), I ended up in the nearest thing to a pub that I could find. Sitting at the bar, sipping a beer and watching a game show on TV, I began to regain some equilibrium. Until the barman, who I had been chatting to on and off, leaned over and said:

'You're not gay, are you?'

'Erm, no. Why?'

'This place becomes a gay singles bar at about this time of the evening, and you might feel a bit uncomfortable.'

'Oh. Oh, right. Thanks for the tip.'

'Actually, I'm finishing my shift. You want to come and play some pool in a different bar?'

This confused the hell out of me all over again. But I went anyway. And either he was straight or he didn't fancy me. We met up with a bunch of students and I beat them all at pool. The affinities between particle physics and games of marbles are well documented, but less well known is the fact that we particle physicists all excel at pool and snooker.

The next day I met a giant of particle physics in the lift.[139] John Collins is one of the people who proved that, in one important sense, quarks and gluons are real.

The proton is made of quarks and gluons. We work out how they are distributed inside the proton mostly by scattering electrons off them – this is what the experiment I worked on at the time in Hamburg was doing. Collins (with two colleagues) had proved that this knowledge 'factorised'.

The way quarks and gluons are distributed in the proton is difficult, maybe impossible, to calculate. But the factorisation theorem tells you that if you measure it in one kind of collision, you can use the information in others. That is, you could use what you learned at our experiment to predict what would happen at different experiments involving protons. Since this is something that we assume, for instance when predicting results of proton collisions at the LHC, it was a very important proof to have.

As to how well he plays pool, I haven't yet found out. I was a bit shy. I think I said 'Hello.' Or possibly, 'What floor is the Human Resources office on?'

So spaced out was I still that it took us more than a day to realise that when I had said 'Business' in response to the official's question about my three-day visit, he had not spotted, or processed, the J-1 visa that would let me work and which was the whole point of the trip. He had instead stamped a B-1 visitor visa in my pristine passport.

Luckily the HR department managed to sort out the missing J-1. Whether the LHC will sort out the missing supersymmetry remains to be seen. In fact, at this stage, the Higgs boson was still missing.

---

[139] Elevator.

7.6 Fun for Some, But Dangerous

The LHC was running well, data were accumulating at quite a rate, and we were glued to the four-lepton and two-photon mass distributions, seeing whether hints of the previous year would be confirmed, or vanish like mist in the new statistics.

There was a collaboration meeting in May where the first update on the two-photon distribution using 2012 data was shown, strictly for internal consumption. I was not working directly on the analysis, and it was very hot off the press. Rumours had been flying around the collaboration, but they changed every day, so it was very hard to get a sense of what was going on. I remember sitting in the audience waiting to see the crucial plot. For some reason, I did not look ahead in the slides. It's possible they had not been uploaded to the agenda page.

In December, with our hints and the CMS hints, I had started to think the odds were better than evens that the Higgs existed. Quite a step, for a confirmed Higgs sceptic. Now, when I saw the first 2012 mass distribution, I knew in my guts that we had it. The data needed checking, and were not really significant enough to be sure, but there was a small bump in the same place as the December results. This felt real. I could not actually, scientifically, be sure. But my stomach was having none of that.

This is a very dangerous phase for a scientist.

The next update on the ongoing Higgs hunt was planned for 4 July. Data were still coming in, cross-checks were still being made. We lapsed into a strange kind of state, desperate to know the answer but desperate to avoid any spoilers, any credible rumours from CMS. The rumour mill had been running for a while, of course. There was even a hashtag higgsrumors (US spelling!) that trended briefly on Twitter. All very entertaining for the neutrals, and it was pleasing that we were not the only ones interested in and excited by our experiment.

But . . . when it came to CMS data, I really did not want to know.

Part of the point of having two independent experiments is that they cross-check each other – independently. We do that most effectively when we are blind to the other experiment's data, right up to the last minute. In fact, up to a point we even try to blind ourselves to our own data. As much as possible of the analysis should be optimised and decided in advance, before looking at the key data. This prevents even the possibility of subconscious bias entering the studies. If you are biased, the truth will probably still out in the end, but in the meantime your statistical estimations of confidence and significance will all be wrong. When your guts think they have the answer, you have to be even more careful. Guts can be wrong.

So hearing gossip from CMS would be at best distracting, a babble of inaccurate noise. At worst, it would be accurate and would bias our analysis. Likewise, as well as betraying confidences and damaging trust within the collaboration, leaking our own ATLAS data could bias CMS.

There was a period back on ZEUS when we had a few more events than we expected, right at the highest energies we could reach. This was terra incognita, no one had done the physics we were doing up there before. So it could have been something really new and exciting. While we were working on this, before we had published our results, rumours flew around that the other experiment on the HERA ring, H1, had also got something weird happening at high energies. They also got rumours about our data. The rumours reinforced each other, lots of people got overexcited. But it was a false alarm, sadly. Once the data got out, it was clear that while we both had anomalies, neither was very significant; worse still, they were not the same. So rather than reinforce each other, our results cancelled each other out.

No great harm was done. Lots of speculative theory papers were written, but that happens anyway. In my first (and, for a very, very long time, only) Radio 4 experience, I was interviewed by Nick Clarke on the *World at One*, which was tremendous. He was a brilliant interviewer. But in the research, some time was wasted, and if we had not gone on to take more data that made it completely clear nothing odd was happening, the result might still be causing confusion today. On the LHC, we didn't want this to happen with the Higgs. We wanted the real answers, as unbiased as possible, as soon as possible. I'm all in favour of scientific openness. In the end it's essential. But *in the end*, not during the experiment. Right then, we really didn't want to know.

The CERN seminar was to be at the start of the International Conference on High Energy Physics (ICHEP). There is one of these every two years. In 2008 in Philadelphia, the LHC had been about to turn on for the first time and the Tevatron had just ruled out their first mass point for the Higgs. Two years later, in Paris 2010, first LHC data were shown and Higgsteria was intense. This time, in Melbourne, would not be the last. But we all knew it would be another big step, one way or another.

We would be surprised at how big a step it would turn out to be.

EIGHT

Discovery

July 2012

8.1 The Announcement

It was 3 July 2012 and I was in Salle Curie, one of the conference rooms below Building 40 at CERN. There are four of these rooms (Andersson, Bohr, Curie, Dirac), and the weekly meetings of the Standard Model group were usually held in Salle Curie. However, this morning Fabiola Gianotti, the spokesperson (meaning boss) of ATLAS, would be rehearsing the talk she would give the following morning. The talk was entitled, with studied neutrality, 'Status of Standard Model Higgs Searches in ATLAS'. It would be given on the morning of the following day, with a webcast around the world and especially to ICHEP, which was just opening in Melbourne.

It had become increasingly clear, initially to us and gradually to the media, that this was likely to be the big one. Peter Higgs had been sighted in town (having lunch with Edinburgh colleagues involved in the search) and François Englert would be in the audience on Wednesday too. There was a definite tension in the air as Fabiola prepared to speak. Very few of us had seen all the ATLAS results collected together. Some of them were only hours old. We all knew we had something special, but how would it stack up? And would Fabiola stick with Comic Sans?

That question was answered immediately. ATLAS and CERN are not very corporate, and no one is going to tell Fabiola how to make slides. Comic Sans it was. The content was more important, of course, though presentation makes a difference and as Patrick Kingsley pointed out in the *Guardian*:

> Comic Sans may be overused, it may look silly, and it may have been designed in a hurry. But it's also very legible, and tests have shown that it makes complex information easier to understand. There's a reason it's used by dyslexia coaches: it facilitates reading.

So, perhaps not a bad choice for communicating tricky new results in as friendly and accessible a way as possible.

The question of how the data would stack up was answered over the next hour. The 2012 run had, up until two weeks earlier, delivered 6.6 inverse femtobarns of data. The recorded data equated to something like 5km of CDs stacked on top of each other, and more than 90 per cent of it had already been analysed and would be included in these results, along with the data from previous years. Several important Standard Model processes had been measured quite precisely, showing that both the detector and the physics were pretty well understood in these collisions.

The key results would be the search for a bump in the two-photon mass distribution and in the four-lepton mass distribution, and Fabiola spent some time discussing the details of the photon identification before showing the result. From the 2011 data alone, we had a 3.5 sigma significance. From the new data alone we had 3.4. The combined data set was 4.5 sigma. Very strong evidence, but not, on its own, up to the conventional 5 sigma threshold to call it a discovery.

But there was more. Fabiola then gave a brief discussion of electron and muon reconstruction, and showed the four-lepton distribution. Another bump. In 2011 it had a significance of 2.3 sigma. In the new data alone, 2.7. Combined, 3.4 sigma. More strong evidence, still not 5 sigma. But put these together, also with the WW and other decay channels from the 2011 data (the 2012 update for those was not yet ready) and the magic number came up. Five sigma.

An arbitrary threshold, just a convention. But one we had set ourselves in advance – we couldn't move the goalposts even if we wanted to be more cautious. We didn't have to beat about the bush. We had a discovery.

This was a very powerful experience. The moment when Fabiola showed our data, and our conclusions, hit me hard. I had seen some of the slides already, and the documentation and the analyses behind them. These were the work of hundreds of colleagues, many of them more directly involved than me in this particular analysis. And years and years of work for us all lay behind the results. But even knowing what was coming, seeing Fabiola declare to all of us what we had done was surprisingly emotional.

At the end of the talk, we decided to stop calling it an 'excess of events' and call it a new boson.

The rest of the day for me was spent in a Meyrin kebab-and-pool hall with Channel 4 News, which is where we came in at the start of the book. I then went to the airport for my semi-regular 20:05 flight to London City Airport. I would not be in Geneva or Melbourne for the announcement, but in Westminster, watching the webcast with lots of UK particle physicists and journalists, the science minister David Willetts, and many other members of the UK science establishment.

We were there rather early in the morning. Jim Virdee from CMS and I were going to talk to the BBC Radio 4 *Today* programme before the talks by Joe Incandela (the CMS spokesperson) and Fabiola. I still didn't know the final CMS result. Rolf Heuer, Director General of CERN, had seen both the ATLAS and CMS results. Did he look excited? Relaxed? Anxious? Worried? Actually, he looked completely inscrutable, as usual. Jim and I were frantically showing each other the results of our respective experiments so we would know what to expect. CMS showed their results first (we had gone first in December) and had 5 sigma if the photon and lepton channels were combined. I was glad I'd had a moment with Jim, because by the time Joe began the first talk we really didn't have much time to actually listen, as there were a lot of people looking for physicists to talk to. I really liked the fact that the seminars were clearly aimed at fellow scientists, not the media or public, but this did mean they had to be translated.

A moment that everyone understood, at which I turned round mid interview and clapped, was the spontaneous round of applause when Fabiola showed our 5 sigma result. Then there was an emotional Peter Higgs meeting François Englert for the first time, and Rolf Heuer's famous declaration that 'As a layman, I would now say – I think we have it. Do you agree?'

We agreed.

After all the rumours and the hints, all the projections and the hows and whys, finally we had, beyond reasonable doubt, discovered something fundamentally new.

Pretty much anything could in principle have turned up at the LHC, since no one had done this before. But if the Higgs boson had not shown up, our understanding of fundamental physics, as encapsulated in the Standard Model, would have been shown to be incomplete. Well, let's be frank, it would have been wrong.

The chain of reasoning is amazing. We knew that the origin of mass occurs at LHC energies. We knew this because two fundamental forces, electromagnetism and the weak nuclear force, unify at these energies. The reason these forces look different to us in everyday, low-energy life is that the force-carrying particles for the weak force, the W and Z, have mass and the photon does not. We had, in the Standard Model, come to the conclusion that this mass can only happen if a certain kind of quantum field fills the universe and sort of sticks to some particles to give them mass. That is indeed quite an extreme leap to make, based on some fairly esoteric mathematics. The only way of proving whether we'd done the right thing or not, whether the field is real or not, was to make a wave, an excitation, in the field. This wave is, or would be, the Higgs boson. And it has to show up at the LHC or the field is either not there or is very different from what we expected. There was nowhere to hide.

Inventing a whole-universe-filling field to make your maths come out right is pretty radical. But it was looking as though it might just have worked. On 4 July 2012, we had seen something fundamentally new, which fitted the description of the particle predicted by mathematical understanding of previous data, coupled with some prejudices about aesthetics, symmetry and how a decent universe ought to hang together. I don't know about you, but this still amazes me.

A nice touch at the end of the event in Westminster was that Alison Boyle from the Science Museum in London got John Womersley (Chief Executive of the Science and Technology Facilities Council, which is responsible for funding particle physics in the UK), Jim Virdee (ex CMS spokesperson) and me to sign a copy of the press release 'for posterity'. It would have been better if David Charlton from Birmingham had signed for ATLAS, since he was Fabiola's deputy (and later became head of ATLAS), but he was busy in Melbourne. And anyway, it was a team effort . . .

8.2 Beyond the Onion: The LHC Computing Grid

I mentioned the huge amount of data we had analysed to get these results. Those data were collected by the different layers of technology surrounding the collision point (tracking detectors, calorimeters, muon systems) and pre-selected by the 'trigger', the high-speed online electronics and computing that made sure the fraction of information we could record contained the most interesting events. They were then reconstructed, as discussed in section 5.1,[140] as the first step of working out what particles had been produced, and in the end leading to the result that Fabiola could show.

The computing power to do this did not all exist at CERN, and still doesn't. There is a worldwide grid of more than 140 computing centres, in 35 or so countries,[141] connected by high-speed networks and serving

[140] 5.1 Why Would a Bump Be a Boson?

[141] The main UK centre, our so-called 'Tier 1', is at the Rutherford Appleton Laboratory at Harwell in Oxfordshire.

the physicists on all the LHC experiments, as well as some other projects around the world too. The data transfer rates are often over 10 gigabytes (the equivalent of two DVDs) per second. When not processing data from LHC, the grid runs simulation programs; the simulated data are also vital for making sense of the real data.

Watching the online monitoring of the data flow and processing on this grid gives a powerful and somewhat eerie sense of the global scale of the LHC project, as data snake from Geneva to North America, Taiwan, Scandinavia, India and elsewhere to end up somewhere on a physicist's laptop and be turned into new information about the fundamental workings of the physical universe.

8.3 Don't Think Twice, It's All Right

The physicists at Westminster talked to a lot of journalists on 4 July. Amid the questions, two recurrent themes emerged that still make me smile, and which are still at play in the ongoing coverage of CERN and the LHC.

Having spent the previous few years hedging, being cautious, talking of hints and probabilities, we were now ready to say we definitely had a discovery. This was the big breakthrough – for us. As I said, even in Fabiola's practice talk we were still using cautious language such as 'excess of events', and only then steeled ourselves to say 'new boson'. It felt great. But it wasn't enough for some of the media. We were prepared to say it looked like the Higgs boson, or some kind of Higgs boson anyway, but we would not say 'We have the Higgs boson!' We'll see exactly what it might take for us to say that in the next section, but it became really clear to me in these conversations that there was still a bit of a communication gap between the people working on the LHC and the people reporting on it. For me to be able to say 'We've definitely found a new boson' was the most exciting thing ever to happen to me (professionally, at least). If it turned out to be the Higgs boson, that would be great. But if it turned out to be something else, that would be in some ways even more exciting. The fact that we did not know, for sure, right then, was irrelevant. What really mattered was that it was not a statistical blip or some other artefact of the analysis. It was real, and it was something utterly new. But communicating all that to someone who wants to hear you go, 'Yay, we found the God particle!' is a bit tricky. Especially when you are overexcited and tired.

And now I have said 'God particle', even if it was in direct quotation marks. Which brings me to the second recurrent theme.

The last of the media engagements I did that day was a BBC World Service radio programme called *World Have Your Say*. This was to be a live discussion of the implications of the discovery. Lovely. They did warn me that 'religious implications' would be part of the discussion, but since there were none, I naively assumed this would be a small part and we could spend most of the hour talking about what had really happened.

The programme began sensibly enough, and we had a good 15 minutes or so discussing what had just happened, what CERN was and the technology behind the experiments. All good. In the studio with me was an engaging and smart guy called Sonny Williamson, a composer who had apparently been doing serial interviews all day as a non-physicist, non-expert big fan of the whole thing who had visited CERN on his holidays. There was also a Welsh Hindu. On the phone were Caitlin Watson from the Institute of Physics and the science writer Marcus Chown. All fine.

I worried a little when Marcus left, with 40 minutes remaining, to be replaced by an unsuspecting cosmologist, Andrew Jaffe, and the presenter said we'd move on to the religious implications.

Sonny said there were none. Caitlin said there were none. Andrew and I said there were none, unless you have the kind of religion that instructs you to disregard evidence, in which case you had many bigger problems than the Higgs boson. The age of the Earth, for example. But in general there was no special religious implication in this discovery.

This wasn't enough. There was a phone call to a Christian minister, who said there were no religious implications. The Hindu guy claimed his religion had sort of predicted it all already, which made the scientists splutter slightly. We got a rabbi, and she said there were no implications. There was a phone call to an imam somewhere. He also said there were no implications. In the end it was a workaday pub discussion of how religion and science talk past each other. The presenter at some point got a bit fed up with the fact that the scientists were being 'too diplomatic', and specifically not being Richard Dawkins. But the religious guys were being reasonable, too, on this occasion (with the possible exception of the quantum Hindu chap). There are some pretty odd belief systems out there, but I haven't come across a religion teaching a dogmatic denial of the existence of a scalar field with non-zero vacuum expectation value, and indeed, the Hindu, the imam, the rabbi and the minister seemed to concur.

So, I wasted the evening of the greatest discovery in my field for decades trying not to start a science-versus-religion war with a bunch of reasonable religious people who were also not trying to start a war. It wasn't so much that it was bad, it was just dull. There were so many more interesting things we could have discussed on that day than whether the mutually incompatible myths of the world's religions were or were not compatible with nature.

The whole thing reminded me of a *Channel 4 News* discussion I was part of when Stephen Hawking and Leonard Mlodinow brought out a book containing the claim that God was no longer necessary. Again, this seemed completely specious, albeit presumably on a par with Leon Lederman's regrettable invention of the name 'God particle' when it comes to enhancing book sales.[142]

My guess is that if you can accommodate evolution, astrophysics and the rest with your religious world view, then theoretical cosmology is unlikely to bother you either. And if you are already denying those things,

---

[142] Yes, I know Lederman claims he wanted to call it *The Goddamn Particle* and blames his publisher for the change. But my publishers wanted to call this book something really silly, and I managed to stop them.

then your faith already has you so nicely cocooned from reality that nothing Hawking says will have any impact. At least the Channel 4 discussion took place on a day when nothing much else was on. The occasion was a new book, not a massive scientific breakthrough, so we weren't really wasting precious time.

A final comment on the relationship between science and religion. Sometime later (in May 2013) I participated in a celebration of the life of Richard Feynman at the Bloomsbury Theatre, organised by Robin Ince. Feynman is by any stretch of the term an iconic figure in science, especially in physics. He won the Nobel Prize with Tomonaga and Schwinger in 1965 for the discovery and formulation of quantum electrodynamics, the first internally consistent and quantum field theory. This theory describes electromagnetism with astonishing precision. He represented this in Feynman diagrams, used throughout particle physics and in this book. He had a childlike curiosity and a way of talking and teaching without being patronising or condescending. He played bongos.

Tom Whyntie and Andrew Pontzen (a particle physicist and an astrophysicist) were doing a turn in the show, and like me they took their copies of the *Feynman Lectures* onstage with them. These were essential reading for me as an undergraduate, and still should be for anyone studying physics now. However, they were written in the early 1960s (around the time Peter Higgs wrote his famous paper, in fact), so while they contain many beautiful insights that easily stand the test of time, they most certainly do not contain all of physics. The Higgs-boson discovery is just one example of the vast amounts of new knowledge acquired since they were written.

Feynman remains an inspiration, but he was never a saint and he certainly didn't encode the whole of physics for us to study and interpret in some process of divine revelation. Physics has moved on, beyond where Feynman could have seen, because of the experiments done since. He is one of the giants upon whose shoulders we stand. As in the quotation from Max Gluckman at the start of this book, 'A science is any discipline in which the fool of this generation can go beyond the point reached by the genius of the last generation.' We've gone beyond Feynman, and this would of course not dismay him in the slightest. He was more aware than most that science has no holy books, that it is a work in progress – the joy of finding things out.

8.4 The Definite Article?

The first conference I went to after the July announcement was another in the annual 'Boost' series, this time organised by Marcel Vos in Valencia. Long Spanish lunches should feature in more conferences. Paella and red wine are excellent aids to discussion, and somewhat to my surprise I found it easier than usual to pay attention in the afternoon talks.

The discovery of the new boson made the meeting even more exciting than the previous year's. One thing our minds were quite focused on was exactly the question that many journalists had been grappling with on 4 July. We were sure we had discovered something, but we insisted on calling it a 'Higgs-like' boson. What would it take to remove the caveat? Why do we not just come out and say we've found 'the Higgs boson'?

One of the properties the new boson definitely shared with the Standard Model Higgs boson was the fact that it decayed very rapidly. Typically, a rapidly decaying particle has several decay options open to it and will 'choose' amongst the options randomly. For the Standard Model Higgs boson, the randomness of this choice is weighted by probabilities that are precisely predicted by the relevant Feynman diagrams calculated in the Standard Model.

This is a common situation in quantum physics: probabilities are predicted, but individual events are not. The decay of a radioactive nucleus shows the same situation: we can measure the 'half-life'. Take the unstable carbon-11 isotope[143] as an example. The half-life is 20 minutes. This means that if I have a number of carbon-11 atoms at some point in

---

[143] See 7.3 Waves.

time, 20 minutes later half of them will have decayed, on average. For small numbers, there will be a significant statistical uncertainty in the time taken for half of them to decay, but the half-life remains the most reliable estimate. If the number of atoms is very large, the half-life is a very precise prediction. But there is nothing in physics that can tell you when an individual carbon-11 nucleus will decay. It may decay immediately, or it may last for ages. All you know is that the chances of it lasting for twice the half-life are ½ x ½ = ¼. Three times, ½ x ½ x ½ = ⅛ . . . And so on. But that is just a probability, not a prediction for the exact decay time.

So as we create more and more Higgs bosons and watch them decay, we can build up a better and better measurement of the probability for each decay mode. These are call 'branching ratios'. Think of a Feynman diagram for a Higgs-boson decay. It is reminiscent of a tree, with the Higgs as the trunk branching into a number of decay products.

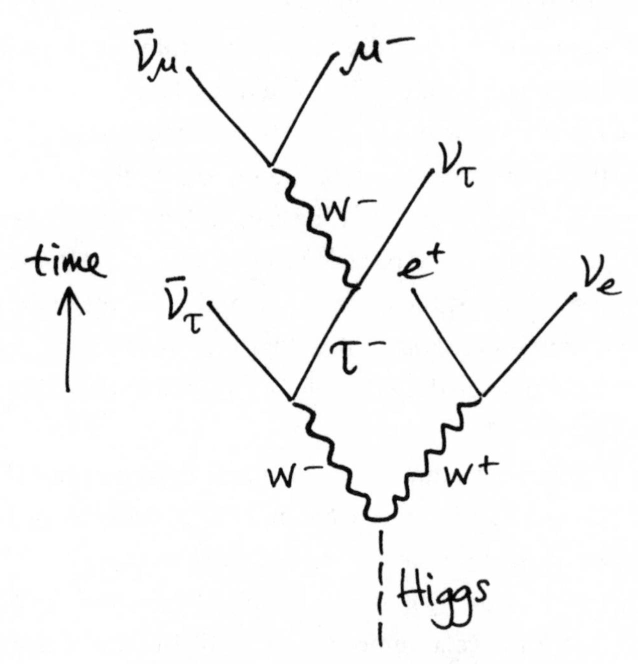

An example of a Higgs boson decay

The branching ratios give us a lot of information about the particle that is decaying. As I said, they are precisely predicted for a Standard Model Higgs boson, at least they are once you know the new boson's mass. This mass is known, from the position of the bumps in the two-photon and four-lepton mass distributions, and is about 125 GeV. Therefore, measuring the branching ratios is a good way of getting some evidence as to whether the new boson we are dealing with is a Standard Model Higgs boson. Do the measured branching ratios agree with the predictions?

The branching ratio measurements have a broader significance, though. In particular, knowing the separate branching ratios to bosons and to fermions is vital. The Higgs in the Standard Model is there to give mass to both bosons and fermions, but the mechanism is very different in each case. Fermion masses are allowed by the presence of this new scalar field – call it the Brout–Englert–Higgs (BEH) field – all over the universe. By coupling to this field, they acquire energy that appears as a mass in the equations governing their behaviour.

The relationship between the gauge bosons and the Higgs is much more intimate than this. When the Higgs field breaks the symmetry between the weak and the electromagnetic forces, by giving the W and the Z mass while leaving the photon massless, something subtle happens to do with the spin of the particles.

The W, the Z and the photon all have a single unit of angular momentum, that is, they have a spin of one. And without the Higgs, they all have mass of zero.

A zero-mass, spin-1 particle has two independent spin states. The spin can point along its direction of motion, or against it. Quantum mechanics tells us there is no other possibility. A boson can be in a mixture of these two quantum states, but whenever you measure it you will find either one or the other.

This is fine for massless bosons, because massless particles always travel at the speed of light. The speed of light is an unattainable maximum speed for a body with mass, so you, as a massive body, can never overtake a massless boson, and indeed you can never even match speed with one. If you could match speeds, the 'direction of motion' of the boson would be undefined, because relative to you it would *have* no motion. But since this can never happen, it is not a problem. The speed of a massless particle is c, to all observers, and so a direction of motion is always defined.

If a formerly massless boson acquires mass, however, there is a problem. Now it is possible for the particle to be at rest, stationary, not moving. Where does its spin point then? To fully describe all the options for a particle at rest with one unit of angular momentum, a third option is needed. Choose a direction and call it 'up', and the spin can point up, down or at right angles to the direction. This happens for atoms with spin-1, for example. We measure three possible spin states. For a moving particle with mass this is the same as saying that the spin can point backwards, forwards or at right angles to the direction of motion.

This may sound a long way from everyday reality, but the fact that the spin-1, massless photon only has two options open to it is observable in the fact that light can be polarised in two (and only two) different ways. This is how Polaroid sunglasses work to reduce glare. When light is reflected from a surface, the amount that is reflected depends upon the polarisation, as well as on the wavelength and the angle at which the light hits the surface. For a range of angles and wavelengths, that polarisation of the light with the electric field oscillating parallel to the surface is reflected more efficiently than the other component.[144] Polaroid material can select one of the two polarisations, cutting out that transverse polarisation and letting the other one through. This has the effect of allowing half the direct light through the glasses, but drastically reducing the glare from reflected light, for example from a wet road. If you take two pieces of a polarising material and put them at right angles to

---

[144] The quantum mechanics behind this is beautifully explained in Feynman's book on QED, which is aimed at a general audience. While hard work, it is well worth a try if you want to better understand quantum physics.

each other, all the light is stopped. There is no third polarisation available to photons.

For a massive spin-1 particle, though, there have to be three possible polarisations. This is what we physicists call another 'degree of freedom'. It is an extra thing needed in order to fully describe a physical system – in this case a boson. Degrees of freedom aren't supposed to suddenly appear like that in fundamental physics. This looks like a big problem for the Standard Model. If the W and Z are to acquire mass, they have to get this extra degree of freedom, and where is that going to come from?

One of the most beautiful things about how the whole BEH mechanism works is the way this problem is solved, at the same time as evading the Goldstone theorem and removing the massless scalar bosons it would predict.[145] The BEH mechanism introduced a new quantum field (the BEH field), which is non-zero everywhere. Because of the type of field it is,[146] it has four possible ways of getting excited, four possible particles, if you like, or four different types of waves it can transmit – four different degrees of freedom. This was a problem for the idea, because three of these four particles would be massless, and none of them had been seen in nature.

The amazing thing is that when these two problems – the extra massless scalar Goldstone bosons and the stubbornly massless gauge bosons with only two degrees of freedom – are put together, they cancel out. Here are the steps:

1. We have four massless bosons – three from the SU(2) bit of the Standard Model and one from the U(1) bit.[147]
2. Of the three SU(2) bosons, one is positively charged, one negatively charged, and one is neutral. The U(1) boson is also neutral.

---

[145] See 4.7 Putting the Higgs in its Place.
[146] A complex scalar doublet if you must know, but you don't have to care.
[147] See Glossary: Gauge Theories (pp.96–9) for the meaning of the U(1) and SU(2) notation.

3. Three of these bosons (the charged ones from SU(2), which are the W^+ and W^-, and a combination of the two neutral bosons that becomes the Z) absorb, or 'eat', the three massless scalar bosons and use them to provide the missing polarisation degree of freedom needed for those bosons to become massive.
4. The photon disdains the food and stays massless.
5. There is one out of the four scalar bosons left. It has mass. That was still a problem, until 4 July, when we found it.

All of this is why we knew the Higgs has to have a mass not too far from the W and Z, and that's why we knew we would find it at the LHC, if it were there. In a sense, it's the same particle, or field, totally mixed up with the W and Z.

That's the intimate connection between gauge bosons and the Higgs. The connection to fermions, on the other hand, is less intimate. The role played by the Higgs here is that it simply allows the fermions to have mass. There's no real prediction for what the masses are, and the way it gives them mass is qualitatively different from the way it gives mass to the W and the Z.

We discovered the new boson through its decays to pairs of photons and pairs of Z bosons, with some indication that it also decays to W bosons. The fact that we saw these decays at roughly the expected rate was strong evidence that whatever we had found, it was connected to electroweak symmetry-breaking. So I was already comfortable calling it a Higgs boson. But we had not yet seen the boson decay to fermions.

The most common decay of a 125 GeV Standard Model Higgs boson is to fermions – to a b-quark–antiquark pair. Unfortunately, at the LHC there are many other ways such pairs can be produced, and the experiments have trouble extracting the signal from background noise. Similar problems applied at the Tevatron, where this decay mode was the best way of looking for the Higgs. This is the problem that the paper I wrote with Adam, Mathieu and Gavin on 'boosted Higgs' searches was meant to address, and one of the reasons the 'Boost' meetings began. In that paper we made some bold claims about how our new technique[148] would have a big effect on Higgs physics at the LHC. One of the things we were discussing over those long lunches in Valencia was whether those claims had been justified.

Our paper pointed out a few ways the b-decays of a Higgs can be picked out from amongst the backgrounds more easily, based around the idea that quite a few of the Higgs bosons produced at the LHC will be travelling quickly – i.e. boosted. The methods we invented work better when the LHC is running at higher energy, closer to its 14 TeV design energy that the 8 TeV we had in 2012. Higher luminosity would also be very helpful. Even so, the paper we wrote had had an effect on the searches in the 8 TeV data – the boost of the Higgs in particular is important – and was cited by both CMS and ATLAS in their Higgs-to-b-quarks search papers. The techniques also inspired a bunch of other papers, some related to the Higgs but some of them about the strong interaction, or about searches for other new physics beyond the Standard Model. Much of that was also under heated discussion in the 'Boost' meetings. I would say the claims had been partly justified so far, but the jury was still out on the full impact of jet substructure on Higgs searches.

Apart from b quarks, the only other fermion decay we had a chance of measuring in these data was the decay to tau leptons. That search was also still going on.

It's a matter of judgement and preference, but if and when we see the Higgs decaying in these two channels – taus and b quarks – at roughly the predicted rates, I will probably start calling this new boson *the* Higgs rather than *a* Higgs. It won't prove it is exactly the Standard Model Higgs boson, of course, and looking for subtle differences will be very interesting. But it will be close enough to justify the definite article.

---

148 See 1.7 Boost One.

8.5 What's in a Name?

Astonishingly, no sooner had a Higgs boson been discovered than some people with access to the media began agitating to change its name. The idea being that naming it after a single person is unfair. This made me unreasonably annoyed.

It's true that many people contributed to the theory behind the masses of fundamental particles. It's true the symmetry-breaking mechanism that gives particles mass should not really be called the 'Higgs mechanism', and I hope I have managed to avoid doing that in this book. The mechanism was proposed independently by several people, two of them (Brout and Englert) predating Higgs slightly, and Hagen, Guralnik and Kibble giving a version that was in some ways more complete, and closer to what is now the Standard Model, but a (very) little later. So it goes.

Even so, Peter Higgs has a claim on the boson itself that no one else has. While its existence was certainly implicit in the other two papers, Higgs' paper was the first to explicitly mention the fact that if the mechanism were to be realised in nature, there would be a new massive scalar boson to find.

For other reasons, an attempt at this point to dispute the name was an embarrassing sideshow, and had a whiff of theorists' arrogance about it. Remember, the theory papers in question date back to 1964. This was before much of the rest of the Standard Model was put together. As the other pieces of the knowledge fell into place – the electroweak and QCD sectors, the discovery of the W, Z and gluon and the full set of quarks and leptons – the mass mechanism became more and more central and the hunt for the missing boson became a higher and higher priority. People started calculating seriously how a Higgs might appear in a collider detector, a very influential and early example being a paper by John Ellis, Mary Gaillard and Dimitri Nanopoulos entitled, significantly, 'A Phenomenological Profile of the Higgs Boson'. Note the name of the boson. That was published in 1975 – only just after first evidence for the Z boson had been seen at CERN in the Gargamelle bubble chamber in 1973, and well before the first real W and Z bosons were observed, also at CERN, ten years later.

Thousands of people contributed to this huge advance in knowledge. The UA1 and UA2 collaborations that discovered the W and Z wrote papers on searching for the Higgs boson, naming it as such. Many of them, on ATLAS and CMS, on the Tevatron and LEP experiments through the nineties and noughties, worked in 'Higgs groups'. Many theorists painstakingly calculated how a Higgs boson would appear in a detector, and they called it a Higgs boson as they did so. Sandra Kortner and Eilam Gross were convening the ATLAS Higgs group when we announced our results on 4 July. All these other people and more had earned a stake in the name by now, and they called it a Higgs. To suggest changing it at this late stage because it looked as though we might actually have found it bordered on insulting.

Of course, it was probably part of the positioning battle now that a Nobel Prize was on the horizon – especially given the arbitrary rule that a maximum of three individuals (not three collaborations) could be awarded the prize in physics. But it was an absurd attempt, at a time when not only were all right-thinking physicists simply celebrating a great piece of new knowledge, but the general public were more aware of a physics result than at any time we could remember, and knew it as a Higgs boson (if we were lucky and they didn't call it a God particle, of course). Not only did Tim Berners-Lee, who invented the World Wide Web at CERN, feature in the Olympics opening ceremony in London that summer, but the Paralympics ceremony included some kind of interpretive dance rendition of the Higgs boson. And called it a Higgs boson. So there.

NINE

What Next?

August 2012 onwards

9.1 I Want an Interstellar Higgs Drive

Life, of course, goes on.

Ten days after the discovery was announced, I was discussing it with Brian Cox in a session chaired by Robin Ince at the Latitude Festival, an annual music and arts festival in the south of England, having earlier recorded an episode of their science chat show *The Infinite Monkey Cage* for BBC radio. I mention this mostly for the bragging rights, of course. Brian may have become used to a world containing rock-star scientists, but I haven't, yet. It is honestly not just for my bragging rights that I mention it, though: the thousands of colleagues who worked on the discovery, as well as the governments and taxpayers who funded it, have bragging rights. So do the thousands (really!) of people who crowded the enormous 'Literary' tent on a Saturday afternoon at a music festival. I think we should all be quite proud of this, the achievement and the public support.

This is evidence, of a sort, that this discovery has had an impact. Impact is a vexed topic amongst researchers, though, and was especially so around this time for various reasons. The 'Science is Vital' campaign[149] had successfully communicated the message that the economic and

---

[149] See 4.1 Science is Vital.

societal impact of scientific research is huge and, indeed, vital. It's a rather utilitarian approach, where the benchmark is 'The greatest happiness of the greatest number' as a measure not just of whether something is worth doing, but of right and wrong. This was the philosophy of Jeremy Bentham, who was very influential in the founding of UCL (and whose dressed skeleton still sits in a glass box in our cloisters), so you might expect me as a UCL person to have some sympathy. And indeed I do.

But other voices in the science community, and broader academia, are frequently raised in objection to this Benthamite approach as a way of judging research. In the UK, academia was gearing up to the 'Research Excellence Framework' (REF), a massive exercise that involved collecting and collating information about research done in all UK universities. As well as looking at academic papers and other 'outputs', this assessment was going to include a look at the 'impact' of research beyond academia. Quite apart from the arguments about the cost of the exercise, the idea of even trying to assess this impact was very controversial.

It is worthwhile unpicking this controversy a bit. We didn't go looking for the Higgs boson because it was useful, or because it would make money; we did it because it was interesting and we are curious people.[150] Do we care about the wider impact or not? Are we trying to have our cake and eat it? Should the success of research be judged by its impact? And what is impact, really? Indeed, quite a few questions at the Latitude event were along the lines of 'What useful stuff can we do with the Higgs now we know it's there?'

A legitimate response is to reply that the sheer wonder of the new knowledge is worth the investment. Part of what it means to be human is to be curious about our surroundings and how they behave, whether you view this as a divinely implanted imperative or a successful evolutionary trait, or both.

It would also be true to say that the technology and training, which are inevitably produced as spin-offs from this kind of research, more than pay

---

[150] Yes, OK, in all senses of the word.

for it. The Institute of Physics is one of the organisations in the UK that points this out most regularly, and in October 2012 MP Alok Sharma hosted a launch event of a new report from the Institute of Physics on 'The Importance of Physics to the UK Economy'.[151] Meeting rooms of the Palace of Westminster can be disconcerting places in which to give speeches: as soon as Frances Saunders, president-elect of the Institute of Physics, began hers, the division bell rang deafeningly for several minutes. When she finally managed to restart, a background rumble from some elderly men in the corner, hitting on the free wine, provided a distracting backdrop. I was told they were lords, which seems too clichéd to really be true. All part of mixing in the corridors of power, I guess.

That Institute of Physics report is full of information about the importance of physics to the UK economy. It defines 'physics-based sectors' of the economy as those where 'the use of physics – in terms of technologies or expertise – is critical to their existence', i.e., if there was no physics, these sectors would not exist. Of course, a similar report could be written about any technologically advanced nation. The use of physics – in terms of technologies and expertise – is critical to the existence of lots of economic activity. To be sure, a physicist would claim that physics is critical to the existence of everything: without it the Earth wouldn't go round the Sun, the atoms of your body would not hang together, and nothing would have mass. But it is possible to delineate a large part of the economy that would quickly be in trouble if we all stopped doing physics research and education.

There are some impressive numbers. The direct contribution to the UK's economic output is £77 billion, or 8.5 per cent. This rises to more than £220 billion if you include indirect spend. Some 3.9 million jobs are supported by these industries in the wider economy, with the average worker in a physics-based industry adding a factor of two more

---

[151] http://www.iop.org/publications/iop/2012/page_58712.html.

'gross value' than the average worker overall.[152] It's not the main reason I do physics, but it is one reason that we should as a nation. And I'm glad someone writes these numbers down, because unless we know these things, physics might stop being done in the UK.

But the question stands and is a good one. What will be the specific impact of the knowledge that this boson exists?

Scientists have many motivations, but one of the fundamental ones is surely a sense of progress: there are things to find out that are important, that are not currently known, and that once you have found them out are added to the body of human knowledge to the eventual benefit of us all. Paul Dirac was a theoretician working on very fundamental physics – relativistic quantum mechanics – and not at all driven by the search for direct applications. However, as recounted in Graham Farmelo's excellent biography, even he cared about impact. He said:

> In my case this article of faith is that the human race will continue to live for ever and will develop and progress without limit. This is an assumption that I must make for my peace of mind. Living is worthwhile if one can contribute in some small way to this endless chain of progress.

Given that, for example, Dirac's equation predicted the existence of antimatter, he certainly contributed.

From the point of view of the arts and humanities, such an idea of progress may appear naive, and certainly it is not obvious that it applies there. Whether human thought genuinely progresses is arguable, and our ability to regress is, sadly, pretty plain. One can even argue about what constitutes progress or regress, and many do. Nevertheless, the recorded history of the human race has a thread running through it of increasing

---

[152] At least two of the successful attendees at the event were ex-students of mine. Gratifying though that was, I was beginning to feel a bit too senior at that stage and was quite glad of the presence of those maybe lords to remind me that I wasn't that far gone yet.

understanding of the universe we live in (including our own bodies), and increasing ability to influence it one way or another. This progress, and the sense of progress, has a big impact on our lives. Our society is built on the fruits of past progress, and we rely on ongoing progress to meet new challenges.

In the UK, as in many other countries, assessment of the 'impact' of research is now embedded in the way funding bodies decide what to fund. This is not only in the REF, but is done by research councils too, who when considering grants for new projects require an 'impact statement', which is a guess as to the future impact of the work beyond academia. You might measure this in terms of patents, jobs created or skills imparted to people, and as I said, it is all controversial. If there is a sliver of Dirac in the heart of every scientist, why so?

One reason is that many groundbreaking applications are serendipitous. The breakthrough comes from an unexpected direction not anticipated at the start of the research and often unrelated to its initial purpose. This is true, though it works in both directions. Breakthroughs in understanding the universe are sometimes unintended consequences of research aimed at direct application. Lots of groundbreaking astronomy, for instance, came about from research directed at improving navigation for trade and the projection of imperial power. Still, it is a fair criticism that guesses as to the future impact of research are likely to miss the most radical benefits, precisely because they will be radically new, unexpected and unpredictable.

A second, much less justifiable reason for controversy is that some academics view the acquisition of knowledge as somehow morally or intellectually superior to the application or dissemination of knowledge. I don't have much sympathy with this one. While a certain detachment from personal gain is morally attractive, I really don't see why striving to understand the origin of life (or mass) is morally superior to striving to cure cancer, for example. Anyway, the fact is that the two activities are very likely to aid each other.

The third reason is that 'impact' assessments are a tool for directing research, and it is not obvious who is best qualified to do this. In fact, many academic researchers resent hugely the idea that research councils or universities try to do this. Like it or not, though, I think someone has to. No single researcher is competent to compare excellence across, say organic chemistry, particle physics, pharmacology and planetary science. So even if, within these areas, peer-reviewed excellence of science were to be the sole criterion, someone has to decide how much resource goes into each area. Unless this is to be done entirely by politicians and civil servants, at least some researchers must engage, sit on committees and argue the balance. And at that level the relative benefits, i.e. impact, of excellent science in each area will inevitably, and correctly, be a factor.

We need a research culture where applications are seized upon and encouraged, not by every single scientist, but by someone in the culture in which they work. Colleagues who take time out of finding new knowledge in order to ensure their work has an impact deserve recognition via the funding system. As a head of a physics department, putting together the evidence for these things was a useful exercise for me and helped the people who do it get recognised. We need both. When she was five, my daughter did a science experiment in school, shining a torch through materials to see which would make the best curtains. Cardboard won. Great science, but poor impact, I fear. I don't think cardboard curtains will rescue our economy.

I'm less convinced that writing predictions of the future impact of all research grants is sensible. In some cases, where research really is directed towards solving a problem with clear applications, perhaps yes. But then presumably you'd write the whole funding case that way anyhow. For other proposals, very often it will be at best a waste of time and at worst it will stifle unexpected breakthroughs and add a short-term bias to research activity and funding decisions. Taxpayer-funded research really ought to be free of such bias, if only to counter the bias that must be present in commercial research.

Hard evidence on this is difficult to collect. I guess ideally one would have to compare two canonical arrays of economies, with one group funding excellent research for its general benefit, and another funding only research with a guaranteed, foreseeable pay-off. I really would not want to be in the second group.

Anyhow, back to the original question: applications of the Higgs. I would be disappointed to think there will never be any, and I think there probably will be some at some point, even though I can't imagine them credibly now. I would not be surprised if I don't live to see them. But if our descendants get an interstellar Higgs drive, or other wild things, they will hopefully thank us for accumulating the basic knowledge behind their amazing new technologies.

And in the meantime, we just have to settle for the spin-offs, and the wonder.

9.2 What Next for the Higgs Boson?

The newly discovered boson can look forward to a period of intense scrutiny, where particle physicists from all over the world measure its properties as precisely as we can to see what secrets it is hiding and whether it will yield any clues that might help us solve some of the remaining puzzles in physics.

Several questions have already been answered. Because we measured the new boson decaying to pairs of Z bosons, and to pairs of photons, we knew immediately it must be a boson itself, that is, it must have an integer angular momentum: 0, 1, 2, etc. This is due to conservation of angular momentum.[153]

In fact, because the photon is massless, we know the new boson can't have spin-1. This comes from the Landau–Yang theorem, which dates back to 1948 when QED was very fresh and people were trying to under-

[153] There is no way to make a half integer (that is a fermionic) spin by combining two integer spins (either two spin-1 Z bosons, or two spin-1 photons).

stand what mesons were. Skip over this if you like, it is quite involved, but it goes something like this:

Firstly, the massless photon has only two possible orientations for its spin, corresponding to the two possible polarisations of light.[154] These orientations are either along or against any direction you choose. We can choose as an axis the direction of travel of one of the photons, so the other photon will be travelling in the opposite direction (at least in the reference frame where the new boson is not moving). There are then two distinct options for how the spins of the photons can be arranged.

One option is that the photon spins both point in the same direction, so their spins add up, giving a magnitude of two for the total spin of the photon pair along that axis. If the new boson only had a spin of one, there is no way to generate a spin of two from it. So we can rule out that option.

We are left with the option that the photons point in opposite directions, so their spins cancel, giving a total spin of zero. This in itself is no problem. The new boson could still be spin-1, since because it is massive, the spin can point perpendicular to the axis we chose. In that case it could decay to two photons that have no net spin along the axis. However, Landau and Yang (working independently) noticed another thing. These photons are identical. The only thing that might distinguish them from each other is the orientation of their spin relative to their own direction of travel. But in this configuration, that orientation is the same – either both photons have spin pointing along their direction of motion, or both have it pointing against. When you have a quantum system and you swap over identical bosons, you get the same system back again, with no relative minus sign.[155] You could do this by rotating the whole system by 180 degrees around the origin, about an axis perpendicular to the axis the photons are travelling along. But for a spherical system such as the

---

[154] See 8.4 The Definite Article?

[155] This is not as obvious as it sounds, and not true for fermions, as discussed in Glossary: Bosons and Fermions (pp.31–3).

new boson, with angular momentum one, rotating by 180 degrees this way introduces a minus sign.[156]

We have finally got there then, because this -1 in the initial state (the new boson) would be incompatible with the +1 required in the final state for the exchange of identical bosons. So a spin-1 boson cannot decay to two photons in this configuration either. If we see a two-photon decay mode (and we do), the new boson cannot have spin-1.

I told you it was involved. But it is something that is said in physics talks very often and not explained, so I wanted to understand it myself well enough to explain it. If you just skipped that bit, welcome back.

More information than this can be extracted from the decays we observe. The angles at which the Zs and the photons are produced contain information on the angular momentum of the new boson, and over the months after the discovery we measured these precisely enough to show that the spin was zero. This doesn't entirely rule out the possibility of a small contribution from other spins, say spin-2, but the main thing we are seeing is definitely a scalar boson.

Another property the Standard Model Higgs boson has to have is that it is 'CP-even'. CP is a combination of two symmetries. The 'C' symmetry (for 'charge') of a system depends on whether it changes when you swap all the additive quantum numbers. This basically means swapping matter for antimatter, so positive charges become negative charges. If the system doesn't change at all then it is said to be even under C symmetry. 'P' is 'parity'. This is easier to grasp: it just means swapping left for right, or inverting the system. This changes the direction of spin, for instance.[157] CP symmetry is almost an exact symmetry of the Standard Model, which

[156] This is true of all odd-numbered spherical harmonics, which are the equations describing waves on a sphere. Think of a wave oscillating on the surface of a sphere, when the northern hemisphere goes up, the southern hemisphere goes down. If you turn the sphere upside down, the phase of the oscillation changes by 180 degrees, or equivalently the amplitude gets multiplied by -1.

[157] Back to the spherical harmonics, all those with even angular momentum (0, 2, 4 . . .) have parity +1, and all those with odd angular momentum (1, 3, 5 . . .) have odd parity, that is, they change sign under inversion.

is a way of saying that physics almost doesn't care whether we are made of matter or antimatter. And yet the observable universe clearly does care – it is overwhelmingly made of matter. To get such an asymmetric universe from such a symmetric theory is a challenge, and requires some source of CP-symmetry violation, which many new theories (such as supersymmetry) can provide. All this boils down to the fact that measuring the properties of the new boson under a CP inversion is an important thing to do, since any difference from the Standard Model might give a clue as to why we are made of matter rather than antimatter.

So CP symmetry depends on what a system looks like when you simultaneously reflect in a mirror and flip all the charges. Since the new boson has spin-0 and no charge, you might think that it is obvious that it does not change under this symmetry, and therefore is CP-even. However, there are examples in nature of scalar bosons with no electric charge that are nevertheless CP-odd. The neutral pion – π_0 is such a particle. It is made of a mixture of up quarks and anti-down quarks, and anti-up quarks and down quarks, combined in such a way that when you swap matter for antimatter (C) nothing changes, but when you reflect in a mirror, the quantum wave function describing the mixture picks up a minus sign. So it is C-even, P-odd, and CP-odd.

The Standard Model Higgs boson is CP-even. The angular distributions of the decay products are also sensitive to the CP of the new particle, and they also indicate quite strongly that, as expected for the Higgs, it is CP-even. Both this and the spin itself will be determined with more confidence as more measurements are made.

As I described in the previous chapter, the Higgs boson can decay in several different ways, and measuring the branching ratio – that is the relative rate – of each of these tells us something new. The decays to the W and the Z are crucial because of the role the Higgs plays in electroweak symmetry-breaking. In addition to providing mass, and giving rise to the Higgs boson, the scalar field of the BEH mechanism provides the extra longitudinal polarisation states that the W and Z need when they acquire mass. Measuring these more precisely will always be a hot topic, since this is such an important and fundamental part of the Standard Model, and there are plenty of options for subtle modifications of the theory.

The decay to photons is a weird one, since because the photon has no mass, it doesn't couple directly to the Higgs boson and is produced via a quantum loop.[158] Any particle that has mass and electric charge could in principle go round this loop – mass is enough to couple it to the Higgs, and charge is enough to couple it to the photon. In the Standard Model, the main contributions are from W bosons and top quarks. But who knows, perhaps particles as yet unseen also contribute? Again, precision measurements of this will always be interesting.

Then there are the fermions: the quarks and the leptons. As already pointed out, picking these out from the backgrounds is a big challenge, but we need to do it because the way the Higgs couples to fermions and gives them mass is completely different from the case of the W and the Z, and so one does not really imply the other.

We can deduce something about the coupling to the top quark because according to the Standard Model, many of the Higgs bosons we see are produced by another quantum loop, this time involving gluons from the proton, which are again massless, coupling to a quark (usually a top quark because it is so heavy) that then couples to a Higgs boson, like this:

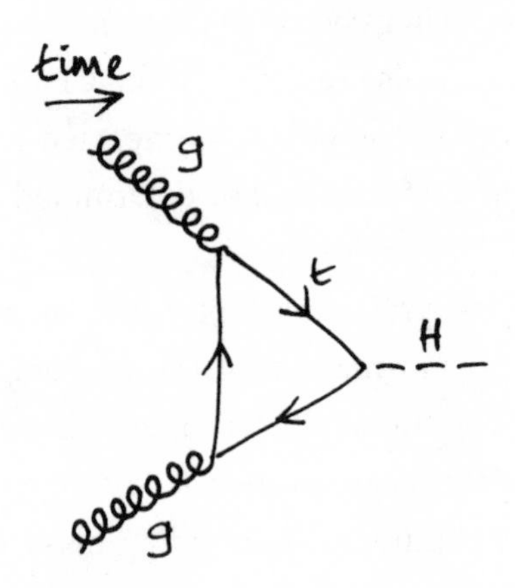

158 See 6.4. Higgs Boson in Crumpling Shocks, Bumps and Stupidity.

So the fact that we see about the same number of Higgs bosons as expected is indirect evidence that the boson couples to the top quark, and thus that the BEH mechanism gives mass to the top quark. However, the top is a bit of an outlier, the heaviest fundamental particle in the Standard Model, and so it is not hard to imagine that something else might be happening with the more 'normal' quarks. Measuring the Higgs decay to b and anti-b quarks is an important test to make, and doing this in the high-energy running will most likely make use of jet substructure in the boosted Higgs events.

Leptons could be different again. The easiest one – because it is heaviest and thus the most likely to be produced – is the tau-lepton. In December 2013, after an exhaustive analysis of the data taken in 2011 and 2012, ATLAS and CMS released compelling evidence (not quite 5 sigma, but well above 3) that the Higgs-to-tau-antitau decay was occurring. Thus the first direct evidence that the BEH mechanism gives mass to leptons. Obviously, making this more significant and measuring it more precisely will be a priority.

All those fermions – tau, bottom and top – are in the 'third generation', containing (along with the tau-neutrino) the heaviest particles. It would be very good to see the Higgs coupling to other generations, to check there is nothing weird going on there. The first generation (up, down, electron, electron-neutrino) is beyond reach for the foreseeable future because their masses are so small that the Higgs decays to them very rarely indeed. But in the second generation (charm, strange, muon, muon-neutrino) we have a shot at the muons and the charm quarks.

For each measurement of a new decay, you can ask whether it really came from the same kind of particle as the other decays. So, repeating the spin and parity measurements for each is important. It is also important to measure the mass. The masses measured in the ZZ and two-photon decays so far are consistent within the experimental uncertainties of a GeV or so, but there is some 'tension', as we particle physicists like to say when we are watching something nervously. The ATLAS measurements are different from each other by about 2 sigma. When the uncertainties shrink and the measurements move around as we get more data, we will be watching very carefully in case it turns out we have two bosons with different masses, one decaying to photons and the other to Z bosons. For the decays to WW and tau-antitau it is difficult, perhaps impossible, to measure the mass with any precision, since invisible neutrinos are produced in tau decays, as well as when W bosons decay to leptons. We should be able to get some measurement of the mass if we can pick out the decay to b quarks, though it will not be as precise as the Zs or photons. If we see it, the two-muon decay channel will have a very precise mass measurement, since muons are precisely measured by both ATLAS and CMS.

If the Higgs mass had been a few GeV lower or higher, we would have had fewer decay channels to measure. Higher, and the Higgs-to-quarks and Higgs-to-lepton channels would have become practically impossible, since the decay to ZZ and WW would dominate once the Higgs mass is high enough to produce real pairs of these particles.[159] A bit lower, and the W and Z decays would have become so rare they would probably be too hard to spot, or at least very much more difficult. As Fabiola said on 4 July: 'If it is the Standard Model Higgs, it's very kind of it to be at that mass.'

Two more things I can think of that I would like to know about this boson.

The decays to the first-generation particles (up, down and electron), all the neutrinos and the strange quark may well be impossible to measure directly. However, it is possible, with the right machine, to measure the total width. This is the real width of the bump in the mass distributions. The width of the bump in mass, or energy, tells us the lifetime of the particle.[160] Unfortunately, in the case of the measured distributions,

---

[159] Twice the W mass is 160 GeV, and twice the Z mass is 182 GeV.

[160] They are inversely proportional to each other. You can think of this as an expression of Heisenberg's uncertainty principle, if you like. The uncertainty in the energy multiplied by the uncertainty in the time is roughly equal to Planck's constant.

the width we see is dominated by the experimental resolution of the detectors. To measure it properly, we probably need a lepton collider, of which more later.

Finally, for the whole Standard Model to hang together, the BEH field must not only give mass to the other particles, but also to the Higgs boson itself. The way it does this involves a 'self-coupling' – Feynman diagrams in which three or four Higgs bosons come together at a point. In fact, pinning this down would allow us to determine the full set of characteristics of the symmetry-breaking field in the BEH mechanism. This is equivalent to a measure of the shape of the bottom of the wine bottle, or if you prefer, the Mexican hat. It is the shape of the background field of the universe. Quite something to know.

Overall, the Higgs boson should have a busy future.

9.3 What Next for Peter Higgs?

When asked how he celebrated the announcement of 4 July, for which he was present at CERN in the auditorium, Peter Higgs replied that he had had a glass of London Pride ale on the aeroplane home from Geneva.

He had more celebrating to come. On 8 October 2013 a substantial fraction of the ATLAS collaboration were clustered around a video screen in Marrakesh waiting for the winners of the Nobel Prize in Physics to be announced.

The will of Alfred Nobel states his estate should fund annual prizes:

> To those who, during the preceding year, shall have conferred the greatest benefit to mankind. The said interest shall be divided into five equal parts, which shall be apportioned as follows: one part to the person who shall have made the most important discovery or invention within the field of physics . . .

(and then goes on to mention four other less interesting fields).[161]

For some reason the Nobel Physics Prize committee doesn't seem to bother too much about the 'preceding year' bit – typically the prize is awarded many years after the discovery. It also stretches 'the person' to mean 'up to three persons'. But it will not award it to more than three and it will not, unlike the Peace Prize committee, treat organisations or collaborations as 'persons'. It also does not award posthumous prizes.

This meant that it seemed very unlikely that anyone who actually discovered the boson experimentally would be in with a shout. It would be impossible to pick out fairly a short enough list of individuals, even if (as some other prestigious prizes have done) the spokespeople of the experiments and the leader of the accelerator were chosen as representatives. The committee was also left with a dilemma in that Hagen, Guralnik and Kibble, the third independent group to publish the mechanism, could not all be squeezed in.

The other dilemma for the Nobel committee was, presumably, at what point could you be sure that the bump in our data, the new boson, was really anything to do with electroweak symmetry-breaking and the origin of mass, and therefore at what point did it constitute evidence that the mechanism proposed by Brout, Englert, Higgs, Hagen, Guralnik and Kibble was actually present in nature? I wasn't party to their deliberations, but for me the combination of the fact that the data said it was most likely a CP-even scalar boson and that it coupled to the Z and W boson,[162] was enough. Probably more than enough. This still didn't mean we could be certain it was 'the Standard Model Higgs Boson'. In a sense we will never be certain of this, since a deviation from the expectations of the Standard Model could show up any time when we measure some property of the particle more precisely than before. In any case, the 'gang of six' did not write down 'the Standard Model Higgs Boson'. The Standard Model did

---

[161] Joke! Really! They are Chemistry, Physiology or Medicine, Literature and Peace. Economics is funded, and awarded, by other means.

[162] The ATLAS and CMS data together constituted very significant signals in both the WW and ZZ decay channels by this stage.

not even exist then, and it would take the work of many theorists and the data from many experiments before it did. The important innovation of the six in the 1960s was to propose a mechanism by which forces based on gauge symmetries could coexist with massive fundamental particles. In the Standard Model, as it eventually emerged, this mechanism was especially relevant to electroweak symmetry-breaking, and therefore the decays to the W and Z bosons were crucial.

The prize went, of course, to François Englert and Peter Higgs, with the third possible slot left empty, which many saw as a fitting tribute to the sadly departed Robert Brout.[163] This raised a cheer in Marrakesh, as did the fact that ATLAS, CMS and the LHC were all acknowledged in the citation.

Prizes, eh? I guess they serve a purpose, and François Englert and Peter Higgs certainly deserved this one. But (and there is a 'but') prizes give a distorted view of how science is done. They encourage the idea that the typical manner of progress in science is the breakthrough of a lone genius. In reality, while lone geniuses and breakthroughs do occur, incremental progress and collaboration are more important in increasing our understanding of nature. Even the theoretical breakthrough behind this prize required a body of incrementally acquired knowledge to which many brilliant people contributed. The discovery of a real Higgs boson, showing that the theoretical ideas are manifested in nature, was thanks to the work of many thousands. There are 3000 or so people on ATLAS, a similar number on CMS, and hundreds who worked on the LHC. While the citation gives handsome credit for all this, part of me still wishes the prizes could have acknowledged it, too. Anyway, perhaps another year. This was a great moment to celebrate, for physics, for particle physics, and for Englert and Higgs.

Higgs, definitely not a man to push himself into the limelight, did a fantastic job, being photographed cheerfully with everyone who asked

---

[163] This is of course why I have been using the term 'Brout–Englert–Higgs' (BEH) mechanism throughout.

(including many of those who worked on the experiments), shaking hands with all comers and, for example, being very visible when the Science Museum opened a big exhibit on the LHC a couple of weeks after the prize announcements. His workload there included a long public question-and-answer session with school pupils, and sharing a platform with the Chancellor of the Exchequer George Osborne, who emphasised the importance of basic research and could be seen as a representative of a long line of UK governments who have consistently backed our membership of CERN.

A nice touch was the fact that Fuller's, the brewers of London Pride, produced a specially labelled edition of 'Outstanding Professor Peter Higgs' bottles of the beer. A bottle (empty) is proudly displayed in my office now.

After a few weeks of this, including the Stockholm trip to get the prize itself, Peter Higgs said he was looking forward to re-retiring.

9.4 What next for the LHC?

High-energy proton–proton collisions in the LHC resumed shortly after the 4 July announcements and continued until 17 December 2012. The additional data reinforced the discovery of the new particle, and allowed us to make some of those more detailed measurements I discussed earlier.

After some running with heavy ions and some technical runs early in the new year, the LHC was switched off for a 'long shutdown'. The message on the status screen said: *End of Run 1. No beam for a while. Access required time estimate: ~2 years.*

There was a big programme of work to complete during these two years. The accelerator would be warmed up from its working temperature of 1.9K (-271 °C) so that people could get to all the connectors of the type that failed so spectacularly in 2008 and test, replace and protect them properly so that the current could be increased to the design value. This will enable the bending magnets to run at full power, and thus allow the energy of the proton beams to be increased towards the 7 TeV per beam, 14 TeV centre-of-mass collision energy that was originally intended.

The detectors are also undergoing extensive maintenance; for example, the entire pixel tracking detector of ATLAS was removed, refurbished and reinserted. The performance improvements and, most importantly, the extra energy will allow us to explore still further beyond the electroweak symmetry-breaking scale, and may throw up surprises. More of that shortly. As I write this, the restart with higher-energy beams for physics is expected early in 2015 – 1 April, in fact.

Looking to the more distant future of the LHC, we will not be able to increase the energy further without completely rebuilding the bending magnets. However, we will get an effective increase in energy, as well as an increase in precision, by increasing the rate of the collisions that the accelerator can provide to the detectors. This is because what really matters is the energy of the quark and gluon collisions. From this perspective, the LHC is a quark-and-gluon collider. The maximum quark or gluon energy we can reach is determined by the energy of the protons multiplied by the fractions of the protons' energy carried by the quarks or gluons. So, if in a given 14 TeV proton–proton collision each quark carried a third of the respective proton's energy, the energy of the quark–quark collisions would be 14 TeV x ⅓ = 4.7 TeV. The chances of finding a quark inside a proton carrying a third of the proton's energy are unfortunately quite small, and as the fraction of energy gets closer and closer to one, the chances drop very quickly. However, it is still true that the more proton–proton collisions you have, the more high-energy quark–quark collisions you will collect, and so the effective energy reach of your measurement is increased.

This is the main motivation behind the LHC luminosity upgrade, proposed to happen in two stages. Phase I is proposed to take place after a couple of years of high-energy running, so sometime in 2017–18 the LHC will shut down again for this. The timescales for Phase II, as well as its scope, are still under discussion. In addition to upgrades to the LHC itself, this will require major upgrades of the detectors so that they can cope with increased data rates. Between them these upgrades would allow the LHC to continue making exciting new physics measurements up to 2030 and beyond.

9.5 What Next for the Standard Model?

One thing we will definitely do with that upgraded LHC, and hopefully with other machines, too, is examine very closely how well the Standard Model works above the electroweak symmetry-breaking scale. This goes beyond the studies of the Higgs boson properties discussed in section 9.2. Remember, this energy regime is qualitatively different from anything we have looked at before. In this regime, the electromagnetic and weak forces are in some sense unified. Certainly their strengths are now comparable. Without the discovery of the Higgs boson, this would have been a no-go area for the Standard Model. The theory would have been unable to make predictions for these energies, and would have been relegated to a low-energy 'effective theory', stunningly accurate for energies below a couple of hundred GeV, but completely out of its depth above the electroweak symmetry-breaking scale.

With the discovery of the Higgs, the Standard Model has a new lease of life. It can make predictions for very high-energy physics – certainly covering everything even an upgraded LHC is able to reach. This is a bold claim, and putting it to the test will be intriguing. One area I find fascinating is the theoretical activity stimulated by the very fact of observing a new boson with a definite mass. A lot of this work is very technical, but one general theme is a re-examination of symmetries and quantum corrections already in the Standard Model to see if they contain more physics than we first thought. There are all kinds of (possibly misleading) clues scattered around and games that can be played. For example, consider a numerical coincidence. The sum of the masses squared of the fermions is very close to the sum of the masses squared of the bosons[164] to within a percentage or so, and given that there are significant uncertainties in the top and Higgs masses, the equality could well be exact. To put it another way, if you had found a symmetry that imposed a condition that the sum of the fermion masses must equal the sum of the boson masses, you could have predicted a Higgs mass of about 123 GeV. Not too far off what we have measured!

The catch is that there is no symmetry we know of that imposes this, so at present it is just a curiosity. There are several numerological games one can play which are at least as plausible. For example, there are three different colors of quark, so should there not be a factor of three in front of those numbers? And there are two W bosons (plus and minus), so why no factor of two? If you did that, you'd get a Higgs mass 'prediction' of 262 GeV. No prizes for that. There are other ways one could make 'predictions' or hunt for coincidences, and the more ways of looking for a coincidence, the less significant a coincidence is if you find it. Look a million different places, and you'll probably find a million-to-one chance turning up. Equally, while a bit of numerology might give a clue, it is only useful if it is a clue to a real dynamical theory. If it doesn't lead anywhere, it is worthless. The way to go is to make measurements and do real calculations, not play number games.

Interestingly, though, there is in fact a similar relation, one that really was derived before the Higgs boson was discovered, and which is based on real Standard-Model calculations. This is to do with the quantum corrections to the Higgs mass that supersymmetry is supposed to help with.[165] Martinus Veltman, who later (1999) won the Nobel Prize with Gerardus 't Hooft for 'elucidating the quantum structure of electroweak physics', worked out what these quantum corrections would be,[166] just considering the known particles – so not including supersymmetry. In

---

[164] That is, $M_{top}^2 + M_{bottom}^2 = M_w^2 + M_z^2 + M_{Higgs}^2$ (all the other particle masses are too small to make much difference). Putting the known numbers in gives: $173^2 + 5^2 \approx 80.4^2 + 91.2^2 + 125^2$, i.e. $29954 \approx 30406$.

[165] See 4.7 Putting the Higgs in its Place.

[166] Acta Phys.Polon. B12 (1981) 437.

fact, since this was 1981, neither the top quark nor the Higgs boson had been discovered, so he was well ahead of his time, and even though he assumed both particles existed, he did not know their masses. Going ahead anyway, he worked out that the first set of corrections (those involving only one quantum loop) would all cancel out if a particular relationship between the particle masses[167] were to hold. Playing around with possibilities, he suggested from this that if the Higgs boson were to be very light, the top mass would have to be about 69 GeV, which back then it could have been. More realistically, if the Higgs mass were to be about the same as the W mass, the top mass would be about 78 GeV.

Unfortunately, when in 1995 the Tevatron experiment finally discovered the top, its mass was much higher, and is now known to be about 173 GeV, and thus the 'Veltman condition' would imply a Higgs mass of about 314 GeV . . . far too high. But all these particle masses pick up quantum corrections and so change with energy, and there are papers being written now suggesting that the Veltman condition or close relatives to it may well play a role in determining the high-energy behaviour of the Standard Model and perhaps avoiding the ugly 'fine-tuning' of the Higgs mass that seems otherwise to be required. Different papers have appeared that claim the Higgs mass is controlled and protected by other approximate symmetries of the Standard Model, such as chiral or 'left–right' symmetry. It is an enthralling spectator sport for us experimentalists, and certainly adds motivation to the efforts to make increasingly precise measurements of key Standard-Model processes.

An important example of these is a set of processes involving the production of pairs of vector bosons – WW, WZ, ZZ especially. The most interesting process here is vector-boson scattering. I talked about this in section 1.2, because without the Higgs this process can't be predicted in the Standard Model. Then I was being a pessimist, suggesting that if the Higgs boson did not show up, vector-boson scattering might be the only place to get a clue as to the mechanism of electroweak symmetry-breaking.

---

[167] $4M_{top}^2 \approx 2M_W^2 + M_Z^2 + M_{Higgs}^2$.

Now we have a Higgs, the process is predicted, and measuring it becomes a very important test of whether the BEH mechanism is really doing the job it is supposed to.

There are plenty of other rare and not-so-rare processes that we can now calculate and which we will be able to measure in the future. We are at the beginning of physics beyond the electroweak symmetry-breaking scale.

9.5 What Next for Supersymmetry and Beyond the Standard Model?

If you believe the headlines, supersymmetry is a particularly tenacious zombie theory that is 'killed', or at least 'maimed' or 'hospitalised', by the LHC every month or two, yet never seems to go away.

There is some truth in this. The LHC data, from the LHCb experiment as well as from ATLAS and CMS (and previous experiments at CERN, too, and HERA and the Tevatron), have ruled out huge swathes of previously possible variants of supersymmetric theories. Yet despite this, as an idea supersymmetry is never likely to go away. The beauty and elegance of the mathematics behind it, coupled with the fact that it is required by string theory, or M-theory, or most likely any other attempt to bring gravity and quantum field theory together, will ensure, I guess, that it remains an important part of the toolbox of theoretical physics, cosmology and mathematics more or less indefinitely.

What is at stake is whether supersymmetry has anything to do with electroweak symmetry-breaking, or with dark matter, or indeed whether it has anything to do with any phenomenon ever likely to be measured in a particle-physics experiment. Does it have any role to play in rescuing the idea of 'naturalness'?[168]

This is very much connected with the loop corrections and mass

---

[168] See 7.5 Is Nature Natural?

relations just discussed in the previous section. Supersymmetric particles also enter those quantum loops, and they guarantee the cancellations the Standard Model on its own seems unable to provide. In doing so, they avoid the fine-tuning problem with the Higgs mass. But fine-tuning is a slippery concept. As well as the LHC and other collider data, precision measurements of the neutron and electron 'electric dipole moments' (essentially a measure of whether the charge distribution of an electron, or a neutron, is spherically symmetric) have constrained the possible parameters of supersymmetry, and to avoid all of these experimental constrains already implies a certain amount of tuning. Most of all, if supersymmetry is to tame the quantum corrections to the Higgs mass, the masses of at least some of the supersymmetric particles should be somewhere close to the electroweak symmetry-breaking scale. If the next run of the LHC does not find anything, this will be become significantly less tenable and many theorists will start to give up on supersymmetry, at least as a solution to the fine-tuning problem. Either way, I doubt 'supersymmetry is just around the corner' will be used as a strong justification for any other large experiment in the future.

Supersymmetry is only (currently) the most popular extension to the Standard Model, however. And cock-a-hoop though the Standard Model may be with its latest success in predicting a fundamental scalar boson and extending its region of applicability well above the electroweak energy scale, the Standard Model is clearly not the full story. There still must be something beyond it, supersymmetry or no.

The most glaring omission is gravity. We have, thanks to Einstein, a very good theory of gravity, but it is not a quantum theory. Space–time is the stage on which quantum field theory plays its part, but at some high energy the idea of a classical space–time comes into conflict with quantum field theory, and we don't know what happens then.

Other problems and omissions include the small point of the missing 85 per cent or so of matter in the universe – the dark matter that is only visible by its gravitational effects on galaxies and other astrophysical objects. Is it a new fundamental particle? It certainly doesn't seem to be explainable by any Standard Model particle. Worse, there is dark energy, which makes up 68 per cent of the stuff (matter plus energy) in the universe. From one point of view, 'dark energy' is just a label for the fact that the rate of expansion of the universe is increasing, for reasons that are unclear. While we are at it, why are we made of matter and not antimatter? And why are there three copies, three generations, of the fundamental particles? And why does the weak force see only particles with left-handed spin, ignoring the right-handed ones? And then what about the neutrinos in all of this, and why are they so light when the top is so heavy? There are a lot of seemingly arbitrary features of nature here that, to a certain type of mind (e.g. mine), plead for a more elegant explanation than 'just because'.

9.6 What next for CERN?

Countries are still joining CERN. In 2012 Cyprus became an associate member. In 2013 Ukraine did likewise, and in 2014 Israel became the first new full member since 1999. Several other countries are in various stages of negotiation to join or to deepen their collaboration. The governments and people of Europe currently show no sign of faltering in their belief that a world-leading particle-physics lab is a good thing to have in the middle of the continent.

Obviously, CERN has its hands rather full upgrading and running the LHC for the next few years. Nevertheless, activities at the laboratory are diverse, with research on new accelerators and new techniques, on antimatter and on nuclear physics, all going on around the site.

These activities include a research effort to design a linear collider, one that could beat the LEP2 record for high-energy electron–positron collisions. Linear colliders have the advantage of avoiding the beam-energy loss due to synchrotron radiation,[169] since the beams do not have

[169] See 1.1 Why So Big?

to bend around corners. However, for the same reason, the beam cannot be stored – the whole beam is lost after one shot and you have to start again. The acceleration then has to be very rapid and the collider has to be very long. The combination of the maximum accelerating gradient you can achieve and the length of the accelerator sets the maximum energy. The only high-energy linear collider built to date was in California, at the Stanford Linear Accelerator Center (SLAC). It contributed to the precision measurements of the Z boson, running concurrently with LEP. Follow-up designs, capable of higher energy, were produced by SLAC and by the Japanese lab, KEK. However, the current favourite is a design that was led by the DESY lab in Hamburg using superconducting accelerating cavities. CERN collaborates in the worldwide effort to construct this machine, and is also conducting research on its own alternative, the Compact Linear Collider – CLIC. Depending on the energy, the superconducting design is several tens of kilometres long. CLIC could shorten this by using a low-energy, high-intensity beam to accelerate a lower-intensity beam to much higher energies.

Another strand is an even bigger circular collider. First surveys have been carried out indicating that it would be possible to build a tunnel for an even bigger circular collider. This would probably collide protons, like the LHC, but would be 80–100km long compared to the 27km of the LHC. The tunnel would extend under Lake Geneva and would surround a couple of mountains, and the maximum energy would dwarf the LHC; the exact value would depend (as with the LHC) on how powerful the bending magnets could be made.

A final example for now is a research project into a really novel method of acceleration – so-called 'plasma wakefield' acceleration. Plasma is a superheated state of matter reminiscent of conditions in the early universe, about 300,000 years after the big bang, in which the temperature is so high that the electrical attraction between electrons and nuclei is insufficient to hold atoms together under the intense rate of high-energy collisions going on. So, electrons and nuclei fly around free, smashing into each other all the time. If a short pulse of a high-intensity beam –

usually a laser – is fired through such a plasma, the charged particles in the plasma oscillate in its wake, and this can lead to very high-voltage differences, as all the positive charges move one way and the negative charges move in the opposite direction. With skill and timing, such a wakefield can be used to accelerate a second beam more rapidly than any other known technique. This has been shown to be a practical way of getting relatively low-energy beams relatively cheaply. The AWAKE ('Advanced Wakefield') project at CERN is trying to do this using a proton beam instead of a laser. Simulations show that in principle if you could fire a beam from the LHC into a suitable plasma, you could get an electron beam with an energy of hundreds of GeV in a few hundred metres, instead of 35km or so for a superconducting linear collider. There are many snags, though, including the fact that the beam intensity would currently be unusably low. It is a speculative, long-term project, and CERN should certainly be doing some of those.

Those are some examples in which the lab is involved. Beyond the Geneva site, though, the council of CERN is charged with 'the organisation and sponsoring of international cooperation' in what the convention calls nuclear research (as of 1953) but is stated elsewhere in the convention as meaning 'research of a pure scientific and fundamental character relating to high-energy particles'. (It also includes cosmic rays.) To address this, as well as the future of the lab, a strategy process was carried out in 2012–13, which fitted in alongside a similar process in the United States and was also strongly influenced by proposals from Japan. The message seems to be that the futures of particle physics and of CERN are inextricably entwined, and you can only really discuss the future of particle physics in a global, rather than a purely European, context. So, on with that, then.

9.7 What Next for Particle Physics?

It was a dark and stormy night. Also, very cold.

Hailstones ricocheted sporadically but fiercely from the ancient stone-framed windows of my large, sparsely furnished room. The walls were also stone, as was the uncarpeted floor. A determined but inadequate space heater gamely plugged away, warming a circular zone about 30cm in radius. I cowered in my big brass bed and struggled to find a WiFi signal. Icy water, the remnant of spent hailstones, ran down the inside of the window frame.

This was January 2013 in Erice, a science centre based in the monastic buildings of a village atop a mountain at the very southwestern edge of Sicily. I was incarcerated there for a week along with the other CERN member-state delegates and representatives of major laboratories and various other partners and organisations from around the world, including from the US and Japan. Our task, before we were released, was to agree on an update of the European Strategy for Particle Physics. There was a lot to update. Quite apart from the Higgs discovery the previous summer (summer seemed so, so far away), the measurement of the neutrino angle θ_{13} was an achievement with profound implications for the future of the field.

The amount of time and money needed to construct a large particle-physics project, and to eventually get science from it, is huge. People who are now too young to have started school will eventually be involved in analysing the data, and many of the people making decisions now won't even live to see the results. So it is important to make a good choice, based on a strategy, and to stick by it, otherwise nothing gets built. The decisions involve many parameters and constraints, and they involve discussions between many well-informed, self-interested parties independently trying to obtain the best outcome for themselves and their cherished sub-field of science. Some of these discussions took place around roaring fires over pasta and wine in the few small restaurants that the organisers had persuaded to open off-season for us. I think this helped. A convivial meal round a fire, with the weather battering the exterior, encourages a certain team spirit. And when the hail and fog were absent, Erice was beautiful, with stunning sunsets and views from the cliffs to encourage long-term vision. I am less sure about the guitars in the Marsala Room, though the Marsala itself definitely helped.

There are the long-term LHC upgrades, which have not yet been approved, or even designed in full. There is a serious Japanese proposal to build a superconducting linear collider, and they were keen to hear what the European attitude to this was – did we want to work on it? Would we contribute to it? There are the neutrinos, and the need for a new 'long-baseline' neutrino facility to explore how neutrinos oscillate. Now that θ_{13} is known to be quite large, there is a real chance we can observe matter – antimatter asymmetry in neutrino oscillations, and thus maybe get a better idea of why we did not all annihilate shortly after the big bang. There are other questions about neutrinos – are they their own antiparticle, for instance?

Returning to higher energies, there are ideas to collide muons. These are heavy versions of electrons, so they have all the advantages of electrons but much less synchrotron radiation (1.6 billion times less, since they are 200 times heavier than electrons). One problem here is they decay in 2.2 microseconds. If you can accelerate them quickly enough, relativistic time dilation helps – time slows down at high speeds and their decay time is therefore much longer. But it's tricky. A nice side effect is that when muons decay they give off neutrinos, so an intense muon beam could be a good way to get an intense neutrino beam almost for free, as it were.

In the end, four large priority projects were identified. I spent two days at the end of the meeting in a freezing-cold room sitting behind a triangular label marked *United Kingdom* while we argued about every single word of a three-page masterpiece.[170] In honour of this and out of respect for my colleagues, I will not presume to paraphrase but will reproduce the four big priorities verbatim here:

---

[170] The full text can be found in the last four pages of this document http://cds.cern.ch/record/1551933, which was signed off formally in Brussels in May 2013. This was in itself a fascinating event, though it's a little vague in my memory due to stopping for 'a couple of pints' of Belgian beer on the way back to the hotel with my fellow delegate.

(c) The discovery of the Higgs boson is the start of a major programme of work to measure this particle's properties with the highest possible precision for testing the validity of the Standard Model and to search for further new physics at the energy frontier. The LHC is in a unique position to pursue this programme. *Europe's top priority should be the exploitation of the full potential of the LHC, including the high-luminosity upgrade of the machine and detectors with a view to collecting ten times more data than in the initial design, by around 2030. This upgrade programme will also provide further exciting opportunities for the study of flavour physics and the quark–gluon plasma.*

(d) To stay at the forefront of particle physics, Europe needs to be in a position to propose an ambitious post-LHC accelerator project at CERN by the time of the next Strategy update, when physics results from the LHC running at 14 TeV will be available. *CERN should undertake design studies for accelerator projects in a global context, with emphasis on proton–proton and electron–positron high-energy frontier machines. These design studies should be coupled to a vigorous accelerator R & D programme, including high-field magnets and high-gradient accelerating structures, in collaboration with national institutes, laboratories and universities worldwide.*

(e) There is a strong scientific case for an electron–positron collider, complementary to the LHC, that can study the properties of the Higgs boson and other particles with unprecedented precision and whose energy can be upgraded. The Technical Design Report of the International Linear Collider (ILC) has been completed, with large European participation. The initiative from the Japanese particle physics community to host the ILC in Japan is most welcome, and European groups are eager to participate. *Europe looks forward to a proposal from Japan to discuss a possible participation.*

(f) Rapid progress in neutrino oscillation physics, with significant European involvement, has established a strong scientific case for a long-baseline neutrino programme exploring CP violation and the mass hierarchy in the neutrino sector. *CERN should develop a neutrino programme to pave the way for a substantial European role in future long-baseline experiments. Europe should explore the possibility of major participation in leading long-baseline neutrino projects in the US and Japan.*

So there you have it. And there were another dozen or so paragraphs on smaller-scale activities, organisational issues and the notorious 'impact'. Obviously particle physics has a lot on its plate.

9.8 What next for Science?

There was surprisingly little backlash in the UK media after the Higgs discovery. I had expected a certain amount, along the lines of 'What a waste of money on something useless that we don't understand.' However, it seems that in Britain at least a pretty large fraction of the public appreciate the significance of the project and to have enjoyed the excitement of the whole thing. It wasn't just a nine-day wonder.[171] Many people had stayed with us from the construction through the initial technical failure, the rumours and hints, the excitement of July 2012 and the eventual Nobel Prize. The fact that the UK winner was the modest and generous Peter Higgs, a Geordie living and working in Scotland, was the icing on the cake. And all that for a couple of quid per person per year. Politically, the economic case for basic research seemed to have made a lot of headway, too. The devil is always in the detailed actions, but the language of many politicians in all the parties about research funding is at least positive at the moment, and I hope and believe the high profile of the LHC has contributed to this. The case has to be remade and re-examined continuously, of course.

---

[171] OK, it did blow up nine days after the first switch-on, but that wasn't the end of the matter . . .

One interesting line of criticism that did show up, though, was that physics had become too divorced from experiment, that experiment was too driven by theory, and that the search for a theory of everything was an obsession that was sustaining a generation of theoreticians who were predicting nothing that could ever be tested against data. This is not a new line of criticism, even within theoretical physics. Lee Smolin of the Perimeter Institute, Ontario, and Peter Woit, of the mathematics department of Columbia University, both published books in 2006 critiquing string theory from this point of view, and Jim Baggott published *Farewell to Reality* hot on the heels of the Higgs discovery, arguing along some similar lines.

There is a reasonable point here, that particle theory probably became too focused on a single track – in this case string theory and its descendants. Such things have happened before. As the *Guardian*'s Ian Sample describes in his book *Massive*, S-matrix theory was all the rage when Brout, Englert, Higgs and the rest were coming up with their ideas, and quantum field theory was something of an unfashionable backwater. They were outside the mainstream, but gradually data drove the mainstream their way. There should always be some people branching out, but it is important they don't all take the same branch – unless the data drive them there. Personally (and it is just a personal preference), I am more interested in the branches that try to solve real observational problems, of which there are many, rather than those that strive for a theory of everything. I wouldn't ban any of them, but I'm a lot more interested in trying to understand this universe than speculating about multiverses.

The LHC is also an illustration of why theorists go wandering sometimes, and why they need to be able to. The Higgs boson was postulated nearly fifty years ago, and the LHC took more than ten years to build – more like twenty if you include the R & D needed before construction could even start. That's a long time to wait. If some people go and do some wild mathematical speculation in the meantime, that doesn't do any harm and may well do a lot of good.

Raising these issues on the back of the Higgs discovery did seem a bit odd. While the existence of the boson had been assumed by some theorists for years, the experimental fact of its existence has fundamentally changed the field, and results from the LHC will continue to do so. The discussion of possible new experiments requires theoretical as well as technical and political insight. Far from losing touch, the political and technical challenges of such machines force us into daily contact with reality, and the data give us access to more of it.

9.9 What Next for Me?

After two years of weekly commuting back and forth between London and Geneva, more or less living on adrenalin and airline food and running a physics group at CERN while teaching at UCL and living in London, I decided that my next managerial role would be in the same city as my family.

So I agreed to become head of the Department of Physics and Astronomy at UCL. This has so far been rather rewarding and means that (counter to the trend for specialisation that pervades much of academia) I have had to broaden my physics knowledge in order to have some clue what the condensed-matter, astro- and atomic physicists in the department are up to.

My research, however, remains centred on the LHC. As I write, Inês, my PhD student, is finishing up the ATLAS Higgs-to-bottom-quark results from the 2012 data, and we have several projects going in preparation for the 2015 restart and the upgrades beyond that. I'll be back.

In the meantime, I also thought it would be worth writing a book about this remarkable period in physics. I hope you enjoyed reading it.

Acknowledgments

I have some people to thank.

Firstly my family, all of them. Thanks for putting up with my frequent absences (mental and physical), and for generally cheering me on or up, as appropriate.

Thanks also to many wonderful friends and colleagues all over the world, several of whom did much more than me to bring about these events: some of you are mentioned in the book, some not. Apologies to those who deserved a mention but didn't get in, and equal apologies to those who are mentioned and might have preferred not to be.

I'd also like to thank UCL and CERN, for the freedom to do this and write about it.

I want to thank Wordpress, Twitter and most especially the *Guardian*, for giving me somewhere to write and an audience to write for.

I owe an enormous debt to that audience, and to many people I have met online, who have encouraged, challenged, educated and amused me, especially through many dull waits in airline lounges. Thank you all.

I'm grateful to many people, including Matthew Wing, Peter Jenni, Albert De Roeck, Bryan Webber, Nikos Konstantinidis, Herbi Dreiner and Nicholas Cizek, for pointing out slips and errors in the first edition. Any remaining problems are still my own fault though.

Thanks to Diane and my publishers, especially Simon (who didn't really suggest a daft title for the book).

And finally, thanks to everyone who pays the taxes, and sustains the society, that allows us to explore the edges of knowledge. It feels quite weird writing something with an author list of less than a few hundred. This book is entirely my fault, but the story of the LHC could have an author list of billions.

Index

About the Author

JON BUTTERWORTH is one of the leading physicists at the Large Hadron Collider and is Head of Physics and Astronomy at University College London. He writes the popular Life & Physics blog for the *Guardian* and has written articles for a range of publications including the *Guardian* and *New Scientist*. He was awarded the Chadwick Medal of the Institute of Physics in 2013 for his pioneering work in high energy particle physics, especially in the understanding of hadronic jets. For the last 13 years, he has divided his time between London and Geneva, Switzerland.